Royal Flying Corps/Royal Air Force Engine Repair Shops France 1914-1918

Aidan J. Williams

ISBN: 978-0-244-62027-1

Copyright © Aidan J. Williams 2017

All rights reserved, including the right to reproduce this book, or portions thereof in any form. No part of this text may be reproduced, transmitted, downloaded, decompiled, reverse engineered or stored in any form or introduced into any information storage and retrieval system, in any form or by any means, whether electronic or mechanical without the express written permission of the author.

Published in Great Britain by

Lodge Books
25 South Back Lane
Bridlington
www.lodgebooks.co.uk

Cover illustrations:

Upper: 130 hp Clerget-powered Sopwith F.1 Camel B3921, Naval 8 Squadron, Royal Naval Air Service, 1917.

Lower: Photographs etc. from author's collection.

Contents

Preface	1
Introduction	5
Pont de l'Arche	9
L F R Fell	15
The Engine Repair Shops	19
Thomas Boland	40
Engines	45
German Aircraft	69
1918 Armistice and Beyond	72
Postscript	83
Endnotes	84
Appendix 1	86
Appendix 2	91
Appendix 3	97
Appendix 4	101
Appendix 5	107
Appendix 6	108
Acknowledgements	111
Bibliography	112

Preface

In the early 1980s, while browsing First World War aviation memorabilia, I purchased a collection of First World War photographs in a militaria shop (now long since disappeared) in Leeds. This collection of obscure photographs was a mystery to me as it related to a unit of the Royal Flying Corps/Royal Air Force in France, the existence of which I was completely unaware: the Engine Repair Shops. I contacted fellow members of 'Cross and Cockade', the First World War Aviation Historical Society, to see if anybody could shed any light on the Engine Repair Shops in Pont de l'Arche, France.

Little or no information was available, so it was suggested that I carry out research into this relatively unknown section of the RFC/RAF myself. I was very fortunate that Lieutenant Commander John Sproule FRAeS (Royal Navy, Retired) wrote to me with details of an interview he had conducted with the Park Commander, Lieutenant Colonel L. F. R. Fell, a few years earlier in 1977. Following correspondence, Lt/Cdr Sproule gave me a copy of the interview on audio cassette. This was the springboard for my research, which culminated in the publication in 1989 and 1992 of two articles in the *Cross and Cockade* journal.

The two articles were originally hand-written then typed up on a typewriter; there was no access to personal computers or the internet at that time. My research involved writing to aviation historians, contacting ex-service organisations, and studying original documents in the Public Record Office (now the National Archives) in Kew, where I discovered much information and data. I met Colonel Fell's son Henry and his wife Catherine; I met ex-Air Mechanic Thomas Boland, who had served in the RFC and had been stationed at the Engine Repair Shops; and I also met ex-Sergeant Donald Winn, who had served as a mechanic in 24 Squadron on the Western Front.

The year 2017 coincides with the hundredth anniversary of the formation of the Women's Army Auxiliary Corps (WAAC); the hundredth anniversary of the award of the Distinguished Service Order (DSO) to Colonel Fell; and also the hundredth anniversary of Thomas Boland's march down Cavendish Street in the centre of Keighley prior to his eventual posting to the Engine Repair Shops. It is, therefore, an appropriate time for the two earlier journal articles to be pulled together, with the addition of newly-researched material, so that the story of the Engine Repair Shops can be presented in a more detailed and readable format to a wider readership.

I received help, encouragement and information from a number of people and sources and this is acknowledged later. All measurements and data are as recorded at the time and, as I do not come from an engineering background myself, I have tried to avoid too much technical jargon in presenting this short history of the Engine Repair Shops, a unit which existed for fewer than five years.

Aidan J. Williams
Bridlington 2017

Take the cylinders out of my kidneys
The connecting rod out of my brain
From the small of my back take the crankshaft
And assemble the engine again

(From 'The Young Aviator Lay Dying', Anonymous. A song heard in the mess of most, if not all, RFC squadrons.)

Introduction

On 28th June 1914, nineteen-year-old Serbian student Gavrilo Princip shot and killed the heir to the throne of the Austro-Hungarian Empire, Archduke Franz Ferdinand, and his wife, in Sarejevo, the capital city of Bosnia and Herzegovina. Due to alliances and international agreements, this was the spark that led to the outbreak of war. A month after the assassination, Austro-Hungary declared war on Serbia; Russia supported Serbia; Germany declared war on Russia and France. On 3rd August, the German army marched through the neutral territory of Belgium to attack France, which left Britain with little choice but to declare war on Germany a day later. Thus began the 'War to End All Wars', which continued until the Armistice on 11th November 1918, culminating in the signing of the Treaty of Versailles in June 1919.

Although the First World War was fought on a number of fronts on land, at sea, and in the air, general interest has always been drawn towards the Western Front and the struggle between the British, French and Belgian forces against the military might of the German Empire. Hard-fought battles, ground won and lost, attacks and counter-attacks, resulted in a line of trenches which stretched from the North Sea to the Swiss Border.

Early Flight and the Royal Flying Corps

On 17th December 1903, Orville Wright made the first controlled flight in a heavier-than-air machine at Kill Devil Hills in North Carolina, U.S.A. It was not until September 1906, however, that the first official flight in Europe was made by the Brazilian Alberto Santos-Dumont. In Britain, it was October 1908 before the first recognised powered flight was made by the flamboyant Samuel Franklin Cody in his Cody (or British Aeroplane) Number One [1]. Nine months later in 1909, Louis Bleriot became the first person to fly across the English Channel in a powered aeroplane, his Bleriot Monoplane.

For a number of years, balloons and man-lifting kites had been used by the British Army for observation purposes and it soon became apparent that powered aircraft could perform a similar function over a wider area, so in 1912, the Royal Flying Corps came into being. The make-up of this new section of the army consisted of a Military Wing, a Naval Wing, and the Central Flying School for pilot training. In July 1914, the Naval Wing became the Royal Naval Air Service and officially part of the Royal Navy, no longer under the jurisdiction of the army.

When war was declared in August 1914, troops and resources were deployed as efficiently as possible given the constraints of the time, and for the relatively new and untested RFC, many logistic problems had to be identified and overcome. Under the command of Brigadier-General Sir David Henderson, the RFC made its way to France with four squadrons (Nos. 2, 3, 4, and 5), an Aircraft Park, and a Headquarters Unit [2].

1913 Royal Flying Corps Military Wing Recruitment Poster.

The expected role of the aircraft in all areas of the war was reconnaissance; not many people had foreseen at this stage the possibility of aerial combat between opposing aircraft. However, as aerial observation and the reporting of enemy troop movements became more and more important in the tactical and strategic planning of engagement with the opposing forces, it soon became obvious that it was vital to prevent enemy aircraft from observing one's own troop movements. Both sides developed ways of limiting the opposition's opportunities for unmolested aerial surveillance, which culminated in aircraft specifically designed to hunt and destroy observation aircraft. The result was that aircraft manufacturers were suddenly tasked with designing scout, or fighter, aircraft, while more resilient and powerful observation and bomber aircraft were fitted with defensive armament for protection.

The RFC Aircraft Parks in France, No. 1 at St. Omer and, by 1915, No. 2 at Candas, played a critical role in supporting offensive operations. The squadrons needed a constant supply of replacement pilots, observers, groundcrew, aircraft, aircraft parts, and engines. Although basic maintenance could be carried out at squadron level, a separate base, at a safe distance from the front line, but with easy communication lines to depots and the Aircraft Parks, was required for more complicated repairs and for the overhaul and thorough testing of all types of aircraft engine in service. The ideal location was discovered at Pont de l'Arche.

Pont de l'Arche

About seventy miles north-west of Paris, and ten miles south-east of Rouen, on the banks of the river Seine (and its tributary, the Eure), near the main railway route to Paris, lies the small medieval town of Pont de l'Arche, which before the First World War was renowned for its footwear industry. At the outbreak of war in 1914, it was a town of fewer than two thousand inhabitants, and it was here, in part of a boot factory owned by the Prieur family (the main factory building survived two world wars and until 2010 was used by Bosch Engineering), that the Royal Flying Corps decided to establish workshops for the repair and overhaul of aircraft engines.

Before it became a separate unit, the workshops formed part of the Army Aircraft Park under the command of Major Allen Douglas Carden, formerly of the Royal Engineers. In 1907, Carden was Assistant Superintendent at the Royal Aircraft Factory (renamed in 1918 the Royal Aircraft Establishment) at Farnborough, where he flew in the 1908 maiden flight of the airship *Nulli Secundus II.* He also flew with S.F. Cody in Cody Biplane No. 1, and later gained his pilot's certificate despite having lost the lower part of his left arm in an earlier flying accident.

Among the officers supporting Major Carden were Lieutenant George Bayard Hynes, who very soon became Depot Commander of the Engine Repair Shops, and Lieutenant Reynell Henry Verney. Lieutenant Verney remained at the Engine Repair Shops (ERS) until December 1915, when he transferred to the Aeronautical Inspection Department (AID) as Inspector of Engines. Because of his knowledge and experience of the Engine Repair Shops, a good working relationship was therefore maintained between the AID and the ERS. After the war he was awarded a permanent commission and eventually retired as Air Commodore.

The AID was formed in December 1913 to inspect all parts and equipment for the RFC and its aircraft, so as to prevent accidents caused by faulty workmanship or poor quality materials. Major J. B. D. Fulton was the Chief Inspector, with Captain R. K. Bagnall-Wild Inspector of Engines, and 2nd Lieutenant Geoffrey de Havilland (before he left in 1914 to join the Aircraft Manufacturing Company, designing many types of aircraft that bore his name) Inspector of Aeroplanes. As the RFC expanded, the AID expanded with it, inspectors being required to offer direction and supervision as well as inspection.

By the end of November 1914, Major Carden had 100 men under his command and the transport section consisted of five trucks, three motorcycles and one Daimler motor car. The establishment of the Engine Repair Shops was soon to grow into a unit fifty times as large, which in March 1918 kept the flying services supplied with engines when the German advance hampered the delivery of supplies from Britain.

Northern France

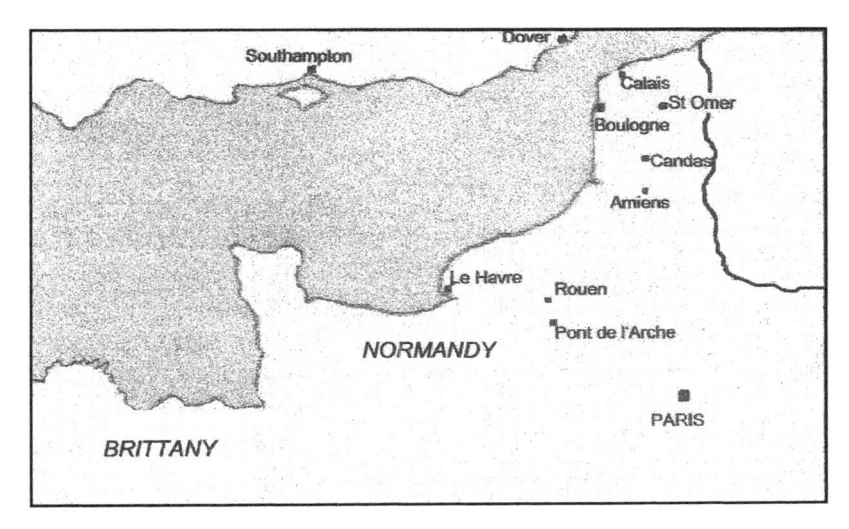

Map of Northern France showing Pont de l'Arche and other important towns, including the Aircraft Depots at St. Omer and Candas. The front line, prior to the Battle of the Somme 1st July 1916, is marked top right.

River Seine

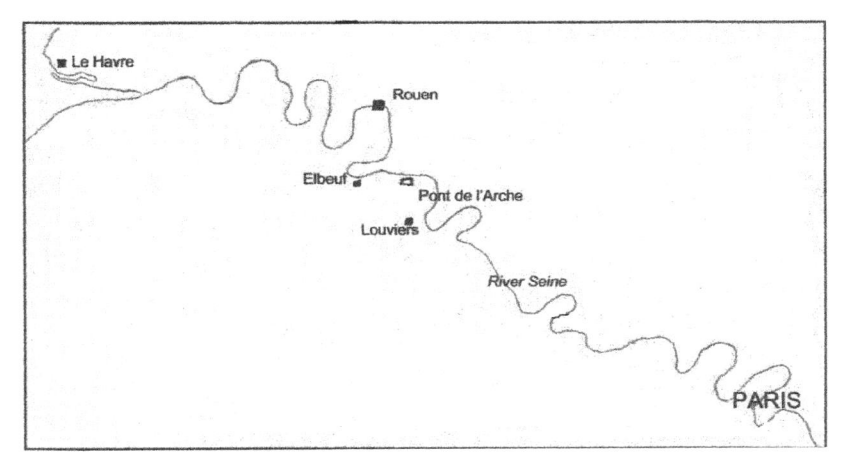

Map of the river Seine showing Pont de l'Arche in relation to Rouen and the Channel port of Le Havre.

By the end of 1915, it had become obvious that the existing support facilities were not large enough to cope with the requirements of the ever-expanding Royal Flying Corps. New Aircraft Parks were formed to supply relevant squadrons: No.1 Army Aircraft Park at Aire; No.2 Army Aircraft Park at Hazebroucke; No.3 Army Aircraft Park at Beauval. St. Omer and Candas were renamed No.1 and No.2 Aircraft Depots respectively, and were responsible for the supply of aircraft and stores to the Parks. Consequently, more land was required at Pont de l'Arche. The sheds on hire to the Engine Repair Shops were therefore put in use as workshops, and the living quarters moved out to a purpose-built hutted camp.

In August 1916, the decision was made by the Army Council to expand ERS even further so as to enable engines from every squadron on the Western Front to be repaired and overhauled with the minimum of delay. The first stage of planned expansion by Brigadier-General Brooke-Popham (Quartermaster General of the British Expeditionary Force, who retired in 1937 as Air Chief Marshall Sir Henry Robert Moore Brooke-Popham, GCVO, KCB, CMG, DSO, AFC) was for the entire boot factory to be taken over by ERS, for which the owner was compensated with the provision of two huts away from the site at a cost of 24000 francs. The establishment was increased to 2000 personnel, and four blocks of engine-erecting shops, tinsmiths' shops, coppersmiths' shops, fuel stores, plus extra living quarters were all completed by the summer of 1917.

The eastern end of the North Camp from the *Avenue Forêt de Bord*. The white-painted trees indicate points of access to the camp (necessary at night with no street lighting). The narrow gauge railway track is the standard British Army 60cm, and the 'V' skip (automatic tipping and righting wagon) was built by either Robert Hudson in Gildersome, West Yorkshire, or by Decauville in France.

Final phase of the expansion of the Engine Repair Shops. Photograph taken from the western end of the South Camp. The road is the *Avenue Forêt de Bord* leading to the centre of Pont de l'Arche in the distance.

L F R Fell

On the 3rd August 1882, Edinburgh-born Colonel William Edwin Fell, of the Duke of York's Own Royal Garrison Artillery, married Alice Pickersgill-Cunliffe from Coulsdon in Surrey. They made their home on the Yorkshire coast in East Ayton, near Scarborough, before moving to nearby Filey. They had nine children, their seventh child and first son was born on 13th January 1892 in East Ayton, named Louis Frederick Rudston Fell.

L. F. R. Fell was educated at the historic St. George's College, Windsor, as a chorister (under the headship of Cork-born Herman Deane). In 1901, as a member of the choir, he sang at Queen Victoria's funeral, and the following year sang at Edward VII's coronation. Fell later progressed to the 350-year-old Tonbridge School in Kent. In 1908 at the age of sixteen, he joined the firm of Clayton and Shuttleworth as an engineering apprentice. At that time, the company had a staff of two thousand workers producing traction engines and steam-powered agricultural machinery.

Fell very soon moved on to the Great Northern Railway (GNR), one of the principal British railway companies which was established under the 1846 London and York Railway Act, and by 1914 he was heavily involved in outdoor railway testing in Doncaster. Following the outbreak of war, Fell enlisted in the Royal Navy on 13th September 1914 and was posted to HMS Pembroke III in Chatham. Identified as an experienced fitter and turner, he was transferred to the Royal Naval Air Service (RNAS) as a mechanic before, at the beginning of 1915, joining the Royal Flying Corps (RFC) at Farnborough. From Farnborough, the five foot ten, grey-eyed, newly-commissioned second lieutenant joined the Engine Repair Shops as Assistant Equipment Officer.

When Lieutenant Fell arrived at the Engine Repair Shops, he found a small unit under the command of Captain George Bayard Hynes, with Captain H. R. Verney as assistant. The Quartermaster was Lieutenant J. Starling (by 1918, Park Commander, Major J. Starling) who was known as the "Old Man", because he was in his mid-forties, compared to the other officers who were in their twenties. Starling had joined the Royal Engineers in 1895 and eventually retired as a Temporary Lieutenant Colonel.

Captain Hynes was born in Malta in 1887 to a naval family and educated in Portsmouth. He was commissioned in the Royal Artillery and qualified as a pilot before being seconded to the Air Battalion in 1911. Hynes remained at ERS into 1919 with the rank of Colonel. By his death in 1938, at the age of fifty-one, he had become Group Captain G. B. Hynes DSO RAF, Deputy Director of the Directorate of Aeronautical Research.

At the time that Lieutenant Fell joined ERS, it was believed that the Engine Repair Shops were large enough to meet any demands which might be made, but when Brigadier-General Brooke-Popham arrived at RFC HQ, he realised the necessity for expansion, and in two years he twice ordered the establishment to be doubled.

In February 1916, Fell was promoted to Captain (Equipment Officer). Hynes, by that time promoted to Major, was commanding what Maurice Baring, assistant to Commander of the RFC Major General Hugh Trenchard, later described as *"one of the most efficient, well-organised, smoothly running and hard-working establishments of the whole war"* (Baring, 1920).

1917 studio portrait of Lieutenant Colonel L. F. R. Fell DSO (H. Fell).

Captain Fell made several visits to RFC Headquarters, and on one of these occasions was invited to dinner with Lieutenant Colonel Moore-Brabazon and General Trenchard. In May 1916, Trenchard had inspected the Engine Repair Shops and both he and his aide, Maurice Baring, were greatly impressed by what they saw. To the majority of personnel at ERS, Trenchard, at the age of forty-five, seemed a comparatively old man, and Fell thought him to be a hard disciplinarian, whilst Baring left Fell with the impression of an amusing university don. On another visit to RFC Headquarters, Fell met aircraft designer and fellow-Yorkshireman Robert Blackburn, who at that time, was attempting to interest the RFC in his 'Skyhook' parachute design; no record of this design has as yet been found, although there is an unrelated safety feature on some modern-day skydiving parachutes which is termed 'Skyhook'.

The Engine Repair Shops

Very quickly, the Engine Repair Shops outgrew the town of Pont de l'Arche to become a completely self-contained community. By the end of hostilities the establishment was five thousand men and women, including five hundred women from the Queen Mary's Army Auxiliary Corps (QMAAC), previously titled the Women's Army Auxiliary Corps (WAAC), and over one hundred officers. The personnel of the camp were housed in rows of wooden sheds and the recently-invented Nissen huts, never far from their place of work. The camp had to provide its own services, such as: food; water; sanitation; electricity; transport; etc.; in addition to the provision of medical and dental facilities.

A gas works was erected, in which used lubricating oil was broken down and processed, to produce gas which could be used in the sheet metal departments and the foundries. There were power-driven hammers in the blacksmiths' shop, and the site could also boast a very well-equipped metallurgical laboratory.

Nissen huts in use as living quarters in the South Camp at Pont de l'Arche. The saw-tooth roofing of the original factory building can just be distinguished in the left background. Note the small marked-out garden areas in front of some of the huts, and washing hanging on improvised washing lines. The row of trees through the centre of the picture marks the main road into Pont de l'Arche from the East, now the D77 *Avenue Forêt de Bord.*

In most establishments, a dynamometer would normally be used to test the power of the engine, the most common example being that made by the Worcester-based company of Heenan and Froude. At ERS however, all engine testing was carried out incorporating either *escargots* (housings within which the fan brake would rotate) or propellers; the engine being mounted on a universal variable air-brake/torque-reaction test bed which was designed by Fell and his colleagues. Fell also designed a variable outlet *escargot* to overcome the need to change fan blades to alter the power. In 1918, Fell submitted an application to the United States patent office, patent number 1369018, in relation to a *'Means for Absorbing Power and Determining the Reaction on Engines'* (see Appendix).

Duration tests were run on full throttle; efficient cooling of the engine was provided by a system of pipes leading from the exhaust ports to a funnel above the *escargot*. The air in the *escargot* was warmed by the exhaust gases; the downward blast on the engine was maintained at a sufficiently high temperature to prevent local cooling of the cylinders, while at the same time preventing overheating of the engine.

There was an iron depositing plant which provided for the plating of worn engine parts; this saved on spares and doubled the production output, particularly of rotary engines. Steel, gun-metal and bronze parts were all treated in the plant; even damaged crankcase cylinder threads on the Le Rhone engines could be rectified by slight depositing. There was little delay in the fitting and machining due to the uniformity and hardness of the deposited surface.

General view of *escargot,* engine test bed, and operator's station.

All the departments were connected by a narrow-gauge railway which led to the purpose-built quay on the banks of the river, where the engines could be received by water from the depot in Rouen and similarly returned. Most of the bogie wagons for this railway were designed and built in-house and were pulled by petrol-driven locomotives.

As the Engine Repair Shops expanded, the new entry of men posted to Pont de l'Arche had little or no mechanical or engineering experience, most skilled workers being retained in reserved occupations in Britain to continue work which was considered vital to the war. To offset this problem, engine repair was divided into three categories: unskilled; semi-skilled; skilled. A trainee would start with unskilled work; if any mechanical aptitude was demonstrated, he would move on to more complicated tasks. Some of the women of the Women's Army Auxiliary Corps, however, had received specialised training on lathe-work and welding, so were considered to be extremely efficient although their number was unfortunately small.

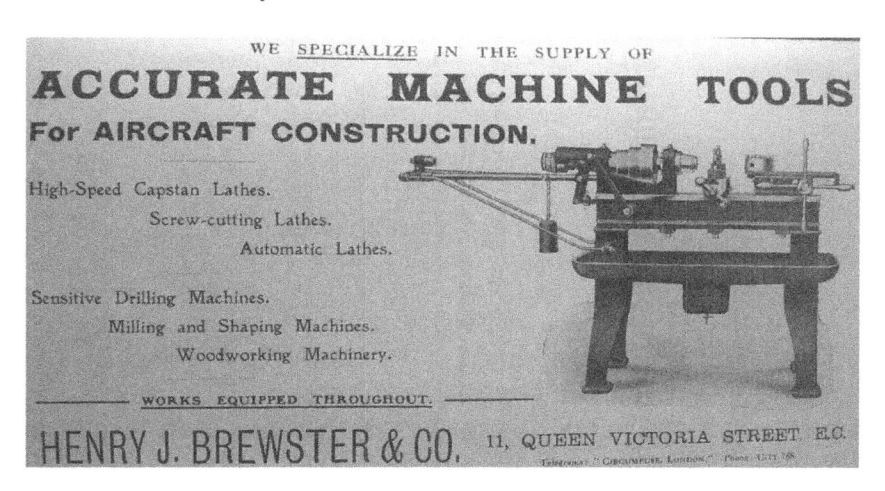

WE SPECIALIZE IN THE SUPPLY OF
ACCURATE MACHINE TOOLS
For AIRCRAFT CONSTRUCTION.

High-Speed Capstan Lathes.
Screw-cutting Lathes.
Automatic Lathes.

Sensitive Drilling Machines.
Milling and Shaping Machines.
Woodworking Machinery.

WORKS EQUIPPED THROUGHOUT.

HENRY J. BREWSTER & CO.   11, QUEEN VICTORIA STREET, E.C.

Contemporary advertisement illustrating a type of lathe similar to those in use at the Engine Repair Shops.

The Women's Army Auxiliary Corps (WAAC) was formed in March 1917 to allow women to carry out war work, thus making more men available for frontline duties. They wore khaki uniform with a detachable collar, white or brown, which helped to identify the different ranks of: Official; Forewoman; Assistant Forewoman; Worker. Coloured inserts on the epaulette designated the different roles of: Driver; Clerical; Mechanical & General; Household.

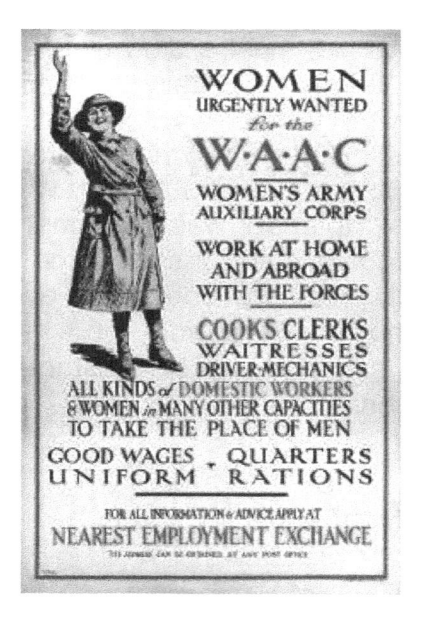

1917 WAAC recruitment poster.

On 1st April 1918, when the Royal Flying Corps and Royal Naval Air Service amalgamated to become the Royal Air Force, the Women's Auxiliary Army Corps was renamed the Queen Mary's Auxiliary Army Corps. Rather than join the newly formed Women's Royal Air Force, female personnel at ERS remained with the QMAAC.

Winifred Wilcox's Queen Mary's Auxiliary Army Corps uniform on display at Seaford Museum in East Sussex (courtesy of Seaford Museum).

In 1906, after the death of her agricultural labourer father, Shropshire-born Winifred Mary Wilcox moved to Liverpool. Having left school, she worked as a dairymaid, farm worker and tram conductor before, after the outbreak of war, becoming a munitions worker for the Liverpool shipping company, Bibby's. In 1917, at the age of twenty, Winifred applied to join the Women's Auxiliary Army Corps, and was very soon training in Hastings before being posted to the Engine Repair Shops.

Winifred was classed as a 'Storeswoman', but she carried out a variety of roles, sometimes assisted by German prisoners. As a storekeeper, dealing with engine spares and used engine parts, she found the work dirty and tiring. In the camp at Pont de l'Arche, she shared a cold hut with nineteen other girls; beds being lined up along one side of the hut, and trestle tables down the opposite side. Their daily routine started with roll-call, a hot mug of tea, then marching to work. There was a roll call in the evening before official 'lights out', which was usually well before 2230 hours.

Winifred's future husband, Arthur Mandeville, son of a retired police officer, was a 'motor tyre repairer' in Brighton until the outbreak of war. In 1914, he joined the Royal Flying Corps as an Air Mechanic, and before long was posted to France. When he first met Winifred in 1918, he was serving as Colonel Hynes's personal driver at Pont de l'Arche. At that time, rations were poor; Winifred recalls hard, maggot-ridden biscuits that could only be broken up with a hammer. The first time they met, Winifred was holding a loaf of bread, which Arthur offered to exchange for a bottle of rum! They met regularly at ERS, eventually marrying in 1920.

Winifred Mary Wilcox (K. Gordon)

One of the machine shops. One of the members of the QMAAC can be seen almost hidden behind the second line of centre lathes. Note the machinery is driven by line shafting and overhead belts, a method that was still in use in some factories up to, and after, the Second World War.

In April 1917, the officer strength at RFC Engine Repair Shops comprised:

Commanding Officer – Lieutenant-Colonel G. B. Hynes
Chief Engineer – Major L. F. R. Fell
Captain E. B. Palmer   2/Lieutenant F. E. Glass
Captain E. M. Bettington   2/Lieutenant H. H. Greig
Captain T. F. Bullen   2/Lieutenant F. Knight
Captain W. D. L. Jupp   2/Lieutenant S. S. Kaye
Lieutenant F. C. Dixon   2/Lieutenant N. Liddall
Lieutenant B. M. Iles   2/Lieutenant T. G.
Lieutenant J. P. Angell   MacKenzie
Lieutenant H. W. McKenna   2/Lieutenant E. P. Proud

2/Lieutenant C. C. Cadman	2/Lieutenant A. W. H. Phillips
2/Lieutenant R. L. Cobb	2/Lieutenant A. S. Poynton
2/Lieutenant J. D. Campion	2/Lieutenant R. T. Royse
2/Lieutenant R. Donald	2/Lieutenant A. C. Smith
	2/Lieutenant L. P. Timmins

The above includes Equipment Officers 1st, 2nd and 3rd Class, whose duties included administrative as well as technical responsibilities.

The Engine Repair Shops was allowed one building (a former community hall) for recreation, which was initially equipped as a cinema. Later this building was enlarged and converted so it could be used as a theatre, and christened 'The Pontodrome'. The theatre became a self-financing project, and thanks to the good number of talented amateur and ex-professional entertainers based in the camp, an excellent and varied programme of plays, pantomimes and concerts guaranteed success at the box office (see Appendix). Theatre takings provided funds for scenery, costumes, sports equipment and film hire. The female workforce at ERS was soon able to take advantage of the provision of a ladies' hairdressing salon, which was equipped and paid for out of theatre performance profits.

The output of repaired and serviced engines continued to increase, from four engines per week in 1915, to 120 engines per week by 1917. The engine types ranged from air-cooled rotary engines, such as the Le Rhone, to water-cooled 'V' engines, such as the Rolls Royce Eagle and Falcon engines, as well as, in the latter stages of the war, the American 400hp Liberty engine. Lighter engines could be physically manhandled into position by three men, but generally, the engines were lifted into the workshops, and onto the test benches, by small hand-operated walking cranes.

In addition to the repairs and overhauls carried out at Pont de l'Arche, many improvements were made to engines that came in. Under wartime conditions there was no time to wait for authorisation from London; consequently, under Fell's guidance, any modifications which were believed necessary were made on the spot, and paperwork (of which there was as little as possible) was completed later.

One of the mechanics posted to the Engine Repair Shops in late 1916 was John Anthony McCudden, the younger brother of James Thomas Byford McCudden VC, DSO & bar, MC & bar, MM, Croix de Guerre. John Anthony joined the Royal Engineers in 1912, and by May 1916 was a dispatch rider. In August 1916, he transferred to the RFC serving at ERS before following in his brother's footsteps and beginning flying training in March 1917. He was first posted to No. 25 Squadron as a Sergeant Pilot before receiving a commission and moving to an S.E.5a Squadron, No. 84. On 18th March 1918, however, Lieutenant John Anthony McCudden was shot down and killed during a dogfight with Manfred von Richtofen's *Jagdgeschwader 1*, most likely by Leutnant Hans Wolff of Jasta 11. His oldest brother, Sergeant Pilot William J. T. McCudden had died in a flying accident in May 1915; Major James T. B. McCudden was killed in a flying accident on 9th July 1918.

Another view of the machine shops, which gives some idea of the size of the former shoe factory building.

American-designed 400 hp Liberty 'V' 12-cylinder engine awaiting test. Towards the end of the war, the Liberty engine powered the D.H.9A and the American version of the D.H.4.

Wolseley Viper engines (a copy of the French Hispano Suiza) ready for testing on the test beds. Towards the rear of the picture, *escargots* have already been wheeled into position.

By late 1917, for administrative purposes, personnel were divided into groups numbered one to six, each group under the command of a Major. The officers, easily recognisable by their compulsory uniform of industrial white coat, took a full part in the work in the machine shops as well as having responsibility for inspection of the work being carried out.

Mechanics working on engine cylinders in one of the workshops at Pont de l'Arche (FlightGlobal).

To keep up morale during working hours, Hynes encouraged competition between the six groups, and Wednesday afternoons were always left free for some type of inter-group sport or games. Occasionally, athletics meetings would be held at the nearby forest ground, where Colonel Hynes would present the prizes. A good example of the professionalism shown by personnel is in the quality of the programmes for the sports events, as well as programmes for the entertainments on offer at the 'Pontodrome', and the production of in-house Christmas cards (see Appendix). Each group's output figures from the workshop would be published, and it was not unusual for men to work extra hours so as to produce higher production figures than a rival group.

Official visitors to ERS were always impressed by the organisation, the high standards, and the obvious skills of the workforce. This technical knowledge and expertise was also recognised by the Air Board, who issued a set of notes to support officers and men of the RFC and RNAS who were to be trained in the intricacies of the aircraft engine. The Preface stated that *"the figures given agree for the most part with standard practice at an important engine repair depot overseas"* (Air Board, 1917); an obvious reference to the Engine Repair Shops at Pont de l'Arche.

Engine manufacturers had not been able to produce accurate figures on the performance of their engines; consequently, the engineers at ERS had to develop their own methods to determine performance capabilities and figures. After the war, and because of his experience with calculating engine data, Colonel Fell, having joined the Air Ministry, established the first national standard for all aircraft engines.

By the time of the Armistice, the Engine Repair Shops had been further extended, taking over much of the neighbouring village of Les Damps. The area is still known by some locals as *Le Camp.* The living quarters were separated into two camps with accommodation for over 5,000 personnel of all ranks. The North Camp consisted of sixty-four huts for RAF and QMAAC personnel; Warrant Officers' quarters and mess; two sergeants' messes; two dining huts; two sick wards; post office; tailors' and shoemakers' shops; various accessory buildings; and the previously mentioned 'Pontodrome'. The officers' quarters and mess were in the South Camp, as were a sergeants' mess; warrant officers' quarters and mess; 130 Nissen huts; dining hall; stores and more accessory buildings.

Workers of the QMAAC in one of the machine rooms at ERS. Note the QMAAC Official, and RAF Officer, wearing white coats. This photograph was taken by official war artist Olive Edis not long after the Armistice. (© IWM Q8116)

Engine Repair Shops personnel. This photograph, dated 22nd October 1918, gives an idea of the numbers of women of the QMAAC ('workers' wearing dark collars, 'forewomen' wearing white collars), compared with the number of male RAF air mechanics.

The actual workshop buildings, which included Braithwaite and Kirk steel sheds, and Bain's sheds, comprised the following:

Building	Size in ft
No. 1 Engine Workshop	133x90
No. 2 Engine Workshop	133x90
No. 3 Engine Workshop	325x70
No. 4 Engine Workshop	325x70
No. 5 Engine Workshop	325x70
No. 6 Engine Workshop	335x70
No. 7 Engine Workshop	335x70
No. 1 Test Shed	148x30
No. 2 Test Shed	145x30
No. 3 Test Shed	148x30
No. 4 Test Shed	227x30
No. 5 Test Shed	174x30
Carpenters/painters shop	174x30
Tinsmiths/coppersmiths shop	133x90
Packing shed and stores	200x32
Unpacking shed	122x30
Packing shed	118x30
Gas and Oil plant shed	60x40
Powerhouse/electrician shop	133x38
Machine shop and fitters shop	200x123
Blacksmith shop and foundry	100x55
Millwrights shop	40x28
MT repair shop	47x25
General offices	210x25
Pump, magnetos/electricians shop	283x30
Engine store, test room, stock room, general stores	200x45

(Institution of Royal Engineers, 1924).

General view of the Blacksmiths' shop.

The drawing office. Because there was limited detail available from the manufacturers, plans for the various engines and engine-parts were drawn up in this office by ERS personnel.

Thomas Boland

In March 1917, one month after his eighteenth birthday, Thomas Boland, a fitter who had served his engineering apprenticeship with Scott Brothers (Lathe Manufacturers) of Keighley, West Riding of Yorkshire, had succeeded in enlisting, despite being in what was classed as a reserved occupation. He and his fellow recruits marched down Cavendish Street in the centre of Keighley to the railway station, from where they travelled to Halifax and were billeted in a tram shed, sleeping on the bare floor.

Having been issued with his uniform in Halifax, Boland was sent to Rugeley Army Camp, Cannock Chase, in Staffordshire, where he completed basic military training. Next he moved to York for a trade test, but when he returned to Rugeley, he discovered his battalion had already been posted overseas. Having passed a second trade test at Charlton Park, Boland was sent to Farnborough, before being posted to Number One School of Technical Training at Wendover in Buckinghamshire, where he spent three months learning the workings of the aircraft engine. Towards the end of 1917, and having successfully completed his initial training, Thomas Boland, now an Air Mechanic in the Royal Flying Corps, was posted to France.

The crossing from Dover was delayed, due to the belief that German submarines were operating in the English Channel, but eventually Boland's ship arrived in Boulogne, only to be bombed by a German airship ten minutes after he had disembarked. He later heard that twenty-eight women from the WAAC had been killed in the raid.

Boland stayed in a camp overlooking Boulogne, not realising how close his tent was to an artillery unit, until the first time one of the large guns was fired! He was transported by train from Boulogne, travelling in the same type of carriage in which thousands of troops had travelled before him, the markings on the side of the carriage read: *'Hommes 40, Chevaux 8',* meaning that the carriage, or more accurately, the wagon, was designed to hold forty men or eight horses. A Crossley tender was waiting for him in Rouen and he was taken from there to his eventual destination, the Engine Repair Shops at Pont de l'Arche.

When Boland arrived at ERS he found that there were separate workshops for each make of 'V' and in-line engine, such as Rolls Royce and Hispano Suiza; but all rotary engines, regardless of manufacturer, were sent to No. 2 Section, to which he was assigned. He worked on Gnome, Clerget and Bentley engines concentrating on the crankcase of the rotary engines which came into his section. His main task was to ensure that the cylinders were perfectly circular and made a gas-tight fit into the crankcase. He achieved this by scraping the inside of the two halves of the crankcase, the semicircular cut-outs would be smeared with blue lead paste and scraped with a file or rubbed on a disc with fine carborundum powder until the shape and size were right. When all cylinders had been fitted and both crankcase halves bolted together, the engine was ready for the balance test, before being prepared for a full engine test in one of the engine test sheds.

Rotary Engine section of Engine Repair Shops, Pont de l'Arche. Mechanics can be seen working on the cylinders and crankcases. Engine parts and tools are laid out on the front benches (FlightGlobal).

Boland's living accommodation was a wooden hut which housed forty men. The normal routine was: *reveille* at 0700 hrs; a cold wash; breakfast; then work began at 0800 hrs. Lunch was 1200 to 1300 hrs, the normal working day finishing around 1700 hrs, six days a week. If anyone was interested in entertainment away from Pont de l'Arche, there was a choice of a walk west to Elbeuf, or south to Louviers.

A 25 centimes note issued by the town of Louviers, south of Pont de l'Arche.

Every Sunday, Boland would assist the padre with Mass in the chapel in the camp, after which he would accompany the padre into the village of Les Damps where Mass was said for the villagers. Before returning to camp, they would go to the nearby Prisoner of War camp and say Mass for the prisoners there. This prison camp, which held about 200 Germans, also provided labour inside ERS, fetching and carrying tools and equipment. Boland would share his monthly tobacco ration of two ounces of tobacco and four packets of cigarettes with a prisoner from Wiesbaden, with whom he became quite friendly.

The chapel at the Engine Repair Shops, where on Sundays Air Mechanic Thomas Boland assisted the padre with Mass.

Engines

It is sometimes forgotten that the aircraft engine was the heaviest and often the most costly element of an aircraft. The early aircraft frames were built from wire-braced wooden parts, and covered with varnished or doped linen; non-vital damaged parts of the airframe could be repaired or replaced relatively quickly by the squadron carpenters and riggers. Damage or failure to some part of the engine very often meant that the only way to return the aircraft to serviceable condition was to replace the entire engine. Some damaged parts could be repaired or replaced in the squadron workshops but, if it was at all salvageable, the damaged engine would be sent to ERS for complete repair, overhaul and testing.

At the outbreak of the First World War, there was still no reliable British aircraft engine available. An aero engine needed to be as light as possible, yet provide enough power to get an aircraft off the ground and keep it in the air. Britain relied heavily on France to provide engines such as Renault, Gnome, Le Rhone, Clerget and Hispano Suiza. The British Royal Aircraft Factory, and private manufacturers such as Beardmore (who produced a copy of the 120 hp 6-cylinder Austro Daimler engine), eventually produced copies of these and other powerplants, until viable British designs appeared, such as Rolls Royce and Bentley.

A contemporary newspaper advertisement for Beardmore engines.

Detail from pen and ink drawing of Royal Aircraft Factory Beardmore-engined F.E.2b (D. Dunne, 1983). Designed as a two-seater fighter to combat the German Fokker Monoplanes of 1915 and 1916, the F.E.2b continued in service to the end of the war being converted to a night bomber with the original 120 hp engine uprated to 160 hp.

Rear three-quarter view of No 46 Reserve Squadron F.E.2d (an improved version of the F.E2b), showing the installation of the 250 hp Rolls Royce Eagle engine (J. Payne).

Aircraft engines in use during the First World War can be divided into three general types: rotary; in-line; and V-type engines. The rotary engines were air-cooled, while in-line and V-type could be either air-cooled or water-cooled.

As the name implies, the rotary engine rotates around a fixed crankshaft; this means that if you were facing the airscrew (propeller) of a rotary-engined machine, you would see the complete engine spinning round with the propeller. Sergeant Percy Butcher, of No. 2 Royal Flying Corps, observes that the rotary engine *"with its cylinders and crank chamber revolving around the crankshaft, with the carburettor supplying the mixture to the pistons through the inlet valve, was really quite straightforward"* (Butcher 1971).

Squadron Leader Cecil Lewis, who was awarded the Military Cross for his work with No. 3 Squadron during the Battle of the Somme, and in 1917 was Flight Commander with No. 56 Squadron, flew many different types of aircraft during the war. He recalls:
"Rotary engines used castor oil as a lubricant … as it was flung out, it burned. And the bitter nutty tang of burnt castor oil is one of the most nostalgic memories of any First World War pilot" (Lewis, 1964).

Samuel D. Heron, however, who while working at the Royal Aircraft Factory was heavily involved in research into air-cooled engines, offers an alternative view:
"The rotary engine, with its fixed crankshaft and rotating cylinders spewing exhaust and castor oil all over the airplane and pilot, is a horrifying device to those used to the modern piston engine airplane" (Heron, 1961).

Lieutenant T. H. Newsome standing next to an 80 hp Gnome-powered Sopwith Pup at a Flying Training School in England.

An in-line engine has the cylinders in one straight line, giving, in some aircraft, a very streamlined appearance, The V-type engine, however, has two banks of cylinders at an angle to each other which, as viewed from the airscrew, give a 'V' outline, thus allowing for a larger more powerful engine with a greater number of cylinders, without the need to extend the length of the aircraft to accommodate an increase in the size of the crankshaft and bearings or a longer row of cylinders.

At one time, during late 1917, the crankshaft grinding machine was out of action at ERS; consequently the crankshafts for some engines including R.A.F.4a, Hispano Suiza, and Beardmore, had to be trued up by hand. This proved to be no problem for the, by now, highly skilled workforce. After careful examination of these crankshafts, it was found that they came well within the laid down limits for parallelism and size.

Benches in the examination area, where dismantled crankshafts are awaiting examination. The notice on the wall at the back is to remind anyone who has been working with cyanide solution to wash their hands before leaving work.

R.A.F.1a Engine

One of the significant modifications introduced at ERS was to the R.A.F.1a engine, a type which was installed in Armstrong Whitworth F.K.3, B.E.2c and B.E.2e aircraft. A problem with this engine was a tendency for the gears to become discoloured and overheated. Fell and his team discovered that this was caused by an airlock in the engine's lubrication system.

The R.A.F.1a was an 8-cylinder 'V' engine incorporating a wet sump and a roller-bearing crankshaft. Oil would come off a flywheel at the rear of the engine and was washed into a duct which went through the roller-bearing in the crankshaft before dropping back in the sump. Investigations at ERS showed that the flow of oil was found to stop at a square-ended joint on the outlet over the gears, which subsequently caused overheating. A solution was devised that permanently cured the problem: a three-sixteenths-of-an-inch hole being accurately drilled at the joint, which then allowed the oil to flow through freely without causing an airlock. This information was passed to the Royal Aircraft Factory, and to the manufacturers of the engine, in order that future engines could be modified before they left the factory. The three-sixteenths-of-an-inch hole was specifically mentioned in the later edition of the *Air Board Engine Book*, where it was stated that the hole had been drilled to prevent "frothing".

Dated 12th October 1915, this photograph shows a line-up of R.A.F.1a-powered B.E.2c aircraft of No.13 Squadron at Gosport, Hampshire, prior to leaving for France and the Western Front (© IWM Q50965). It was at Gosport at the end of 1916 that Major Robert Smith-Barry introduced the 'Gosport' system of flying training, incorporating skilled instructors; specific standards which pilots had to achieve; and utilising the Avro 504 as the basic training aircraft.

R.A.F.1a Lubrication Diagram

AIR BOARD ENGINE BOOK.

(91.)

R.A.F. 1a.   LUBRICATION.

Labels on figure:
- Lip. D.
- Oil Chamber. F.
- Oil Channel. H.
- Inspection plug. L.
- Oil Gallery. E.
- Fly-wheel. A.
- Bye-pass. G.
- Inspection plug. M.
- Oil Gallery Orifice.
- Inspection plug. M.
- Fly-wheel. A.
- Gauze screen. B.
- ¾" dia. Holes. C.
- Oil Leads. J.
- Collector Rings. K.
- Inspection plug. L.

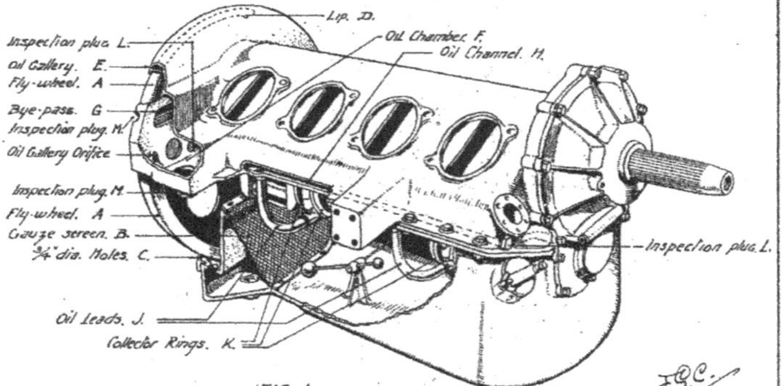

FIG. 1.

In the R.A.F. 1a. Engine no oil pump is used; the flywheel (A) in the crank-case being utilised to convey the oil to the lubrication system.

The oil is poured into the crankcase and it passes thence through the gauze baffle screen (B) into the sump.

The flywheel (A), rotating in its housing, extends almost to the bottom of this sump and drawing the oil therefrom through three ¾" holes (C) in the bottom of the housing, conveys it on its rim to its highest point where a knife-edge lip (D), cast in the top of the housing and just clearing the rim of the wheel, causes the oil to enter the passage or gallery (E) (also cast in the casing) through which it flows into a chamber or reservoir (F) formed in the engine-bearer foot.

The amount of oil in this reservoir is regulated by means of a bye-pass (G) situated so as to provide for 8" head of oil, the surplus flowing through the bye-pass and returning to the sump.

From this reservoir a longitudinal channel (H), cast in the side of the crankcase, conducts the oil to the camshaft driving gears in the front of the engine. In order to prevent frothing, a vent-hole (3/16" dia.) is drilled in the end of the channel.

At two points in this channel connections are made for the ¾" pipes (J) which lead to two crankshaft main roller-bearing caps.

Before entering these caps, the oil passes through plates or dia-phragms, each drilled with a 5.5 m/m. hole which regulates the amount of oil delivered.

Continued on Sk. N° 92.

R.K.M.

ISSUE NUMBER	1.			
DATE OF ISSUE	Jan. 1917			

Lieutenant T. H. Newsome, having been awarded his wings (i.e. qualified as a pilot) is seen standing beside an R.A.F.1a-powered B.E.2 aircraft. The curved air-scoop over the engine can be clearly seen; this provided an entry-point for the air which cooled the cylinders (G. S. Leslie).

R.A.F.4a Engine

In general the R.A.F.4a, which powered the R.E.7, R.E.8 and B.E.12 aircraft, performed well and was considered a reliable engine.

Test for the R.A.F.4a at ERS:
- 60 minutes light run on the *escargot*
- 10 minutes run at 1800 rpm
- Further 60 minutes trial at 1800 rpm to check oil consumption
- Momentarily opened out to 2000 rpm to simulate 100 mph air speed
- Complete dismantling of the engine, apart from the crankshaft and big ends, for inspection
- Final 3 minutes test with propeller, manifolds and all attachments fitted

Hispano Suiza Engine

Another important modification with which Major Fell was involved was improvements to the Hispano Suiza engine, which powered British S.E.5a, French S.P.A.D. fighters, and later the Sopwith Dolphin. High compression pistons were cast in the ERS light alloy foundry [3], with the cylinders ground further up the bore, which increased the compression within the cylinder from 5:1 to 5.6:1. These new pistons were fitted to every Hispano Suiza engine that arrived at Pont de l'Arche. The resultant increase in aircraft performance brought about by the new pistons was considered so vital to the war effort that Fell was awarded the Distinguished Service Order (DSO) for his work.

Hispano Suiza engine fitted to a well-worn S.E.5a.

Test for the Hispano Suiza at ERS:
- Run in *escargot* with fan brake for 5 and 10 minute periods at 900 rpm
- 5 minute run at 1350 rpm
- 10 minutes at full throttle to 1800 rpm
- Removal of oil sump, crankshaft and camcase covers for inspection
- Remake all joints
- 20 minutes at 1800 rpm with minimum oil pressure of 70lbs

In October 1917, the test on a 180 hp Hispano Suiza engine gave the following power curve:

Revolutions per minute (rpm)	Horse power (hp)
1300	136
1400	149
1500 (unmodified carburettor)	151
1600	175
1700	187
1800	198
1900	208

To accommodate the new light alloy pistons, modified carburettors had to be fitted, which were not always being assembled correctly. Major Fell made two or three visits to operational airfields to assess the performance of the improved Hispano Suiza engines and, although in his opinion no carburration was particularly efficient, the modifications appeared to give improved performance at altitude, which was the main requirement at that time.

While staying at these airfields, Major Fell was shocked by the obvious strain under which the flying crews were carrying out their various tasks, which highlighted to him how fortunate were the personnel at the Engine Repair Shops, being stationed at a distance from the rigours and dangers of the front line.

Rotary Engines

Many innovative modifications were made to engines as they arrived for complete overhaul, such as improving the valve seating on the Clerget rotary engine. The coned exhaust valve seats were constantly found to be warped, so at ERS the valve seat was screwed onto the outside flange of the head of the cylinder thus solving the problem. The whole of the cam box on the Clerget was given a watch-like finish which, it was believed, led to a great deal of extra care being taken when the engine was received at a squadron.

Squadron Commander Christopher Draper of Naval 8 Squadron RNAS observed that pilots' knowledge of engines was generally not very good, so in November 1917, he created a list of instructions to assist his pilots in being more sympathetic to the workings of the Clerget engine with which their Sopwith Camels were powered at that time. His list consisted of fifteen 'DON'TS', three examples of which are:

"DON'T open up STRAIGHT AWAY, It does not give the oil a chance to circulate, and ruins the obdurators.
DON'T exceed 1,250 revolutions at any time. It causes the ball-races to 'creep', and other unpleasant things.
DON'T forget that SYMPATHY and a thorough knowledge of all 'work' especially 'carburration' is very important."
(Johnstone, 1931).

The 'obdurators' to which Commander Draper referred (more usually spelled 'obturator', or alternatively *'obturateur'*), were piston rings of 'L' section, designed to ensure gas-tightness by compensating for any distortion of the steel cylinders which might occur due to high temperatures and uneven cooling.

Sopwith F.1 Camel with 130 hp Clerget engine, similar to the aircraft flown by Commander Draper's Naval 8 Squadron RNAS. (© IWM Q56841).

130 hp Clerget 9b rotary engine fitted to the Sopwith Triplane built by the Northern Aeroplane Workshops for the Shuttleworth Collection. In the first view the cam box and nose-plate have been removed to display the ball race and piston connecting rods. The second view shows the same engine, at a later date, with all fittings in place including the propeller and the valve tappet rods (two separate tappet rods and rocker arms operate the inlet and exhaust valves for each cylinder).

Rotary engines would be balance tested prior to a full engine test. The engine was placed on a spigot, where it could rotate freely, while the inspecting officer would gently spin the engine, checking that when it stopped there was no movement backwards or forwards. Having passed the balance test, the engine would be reassembled, the pistons having been fitted with larger piston rings than the originals. Valves, etc. would be fitted, and the timing would be set, before the complete engine test was carried out in one of the test sheds. The test sheds had to be well-ventilated to carry off the exhaust gases, which meant that the constant roar of engines on test could be heard all day throughout the Pont de l'Arche area. The repaired engines would then be despatched to Rouen and from there to either No. 1 Aircraft Depot at St Omer, or No. 2 Aircraft Depot at Candas.

According to No. 24 Squadron mechanic Donald Winn, an experienced groundcrew could change a Gnome engine on a D.H.2 aircraft in twenty-five minutes. The D.H.2 was a pusher fighter aircraft; i.e. the engine, propeller, and ancillary equipment were situated behind the pilot, who sat in an open nacelle or cockpit in front of the wings, thus providing the pilot with a clear field of vision ahead and to both sides. With the engine behind, the aircraft was pushed through the air, rather than pulled by an engine at the front of the aircraft. To fit a replacement engine, Winn would lay along the top of the nacelle, guiding the engine onto the crankshaft, while another mechanic took most of the weight of the engine.

Sergeant Donald Winn, front right, with fellow 24 Squadron Sergeants. All five are wearing RFC uniform including the tunic which was familiarly known as the 'maternity jacket'.

When swinging the propeller to start the engine, the thumb had to be alongside the forefinger to prevent the thumb from being wrenched out of its socket. Winn was always careful not to step too far back after the propeller had been swung, as there was always the possibility of bouncing back from the cross-wires into the path of the propeller. After the engine had started, he would perform a somersault between the upper and lower booms to ensure he was clear of the rotating engine and propeller.

De Havilland D.H.2 with 100 hp Gnome Monosoupape engine (one of a batch of one hundred D.H.2s built by the Aircraft Manufacturing Company). Note how awkward it would be for a mechanic to climb between the booms to access the propeller and then extricate himself once the engine had fired into life. Donald Winn's novel method of avoiding the rotating propeller is described above (© IWM Q57627).

The 110 hp Le Rhone rotary engine posed few problems for ERS, although towards the end of 1917, there occurred quite a number of instances of valve rocker 'T' pieces breaking; this was attributed to the use of unsatisfactory material at the manufacturing stage. During repair, all cylinders were tested to 60lbs air pressure while the valves were tapped, and tested with paraffin for leaks.

Test for the Le Rhone:
- 30 minutes run in *escargot* at 800 to 900 rpm
- 30 minutes full out
- Removal of noseplate and induction pipes for a general internal examination
- 5 minutes full out with propeller

The length of time an engine would last before requiring an overhaul varied for each type of engine; for example the Air Board Technical Notes optimistically stated that:
"The average rotary engine must be taken down after about 40 hours running whilst most stationary engines can be relied upon to run some 100 hours without overhaul". (Controller – Technical Department, 1917).

Rolls Royce engines were generally considered to be capable of 100 hours running time. The Bristol F2b fighter of No. 22 Squadron's Lieutenant T. H. Newsome provides a remarkable example of reliability. The Rolls Royce Falcon engine of Newsome's aircraft, serial number C4706, is recorded as having had 163 hours running time before overhaul, while the replacement engine had 110 hours before its removal for overhaul. Born in 1895, Lieutenant Thomas Henry Newsome enlisted as a private in the Royal Fusiliers, before eventually being commissioned in 1917 when he joined the RFC, becoming a successful pilot with No. 22 squadron in 1918.

ERS mechanics working on Rolls Royce engines in one of the workshops (FlightGlobal).

In practice, as new engines arrived on the Western Front with improved reliability, other factors had to be considered. In the following table, for example, the low figure in February for the Clerget engine was due in no small measure to the effects of severe frost affecting the lubrication system.

Summary of Monthly Returns from BEF Regarding Aeroplane Engines 1917

Average Hours Run Between Complete Overhauls

	250 hp Rolls Royce	160 hp Beardmore	140 hp RAF	90 hp RAF	100 hp Mono (Fr)	100 hp Mono (E)	110 hp Clerget	110 hp Le Rhone	80 hp Le Rhone	150 hp Hispano
February	-	31.44	25.28	79.20	49.05	37.14	11.35	18.16	21.23	-
March	-	23.28	17.34	70.24	28.02	35.57	19.32	24.40	33.52	-
April	-	41.58	25.56	79.08	25.30	31.15	21.36	18.46	34.02	-
May	-	37.51	43.14	92.52	-	-	18.00	23.35	20.41	21.43
June	-	37.56	31.39	78.04	-	-	23.47*	32.55	34.16	38.53
July	30.33	41.00	27.54	55.59	-	-	25.41*	27.23	42.34	29.58
August	15.57	37.30	49.39	112.56	-	-	29.02*	38.41	46.57	44.50**
Totals	46.30	251.57	221.24	568.43	102.37	104.13	149.13	184.16	233.45	135.24
Average no hrs	23.15	36.00	31.38	81.15	34.12	34.49	21.19	26.19	33.24	33.51

*130 hp Clerget
**150/180hp Hispano

66

Summary of Monthly Returns from BEF Regarding Aeroplane Engines 1917 (continued)

Summary of Causes of Failure

	250 hp Rolls Royce	160 hp Beardmore	140 hp RAF	90 hp RAF	100 hp Mono (Fr)	100 hp Mono (E)	110 hp Clerget	110 hp Le Rhone	80 hp Le Rhone	150 hp Hispano	Totals
Breakages & Undue Wear	1	53	4	67	2	9	22	30	9	12	209
Ignition	-	-	1	-	-	-	3	5	3	1	13
Carburration	1	1	1	13	-	1	-	1	-	-	18
Cooling System	1	29	-	-	-	-	-	-	-	1	31
Lubrication	-	3	15	61	16	43	17	20	10	7	192
Hostile Action	2	16	7	13	1	9	10	11	3	12	84
Crashes	2	28	121	104	4	15	97	181	90	55	697
Time Expired	-	-	5	15	8	8	3	27	24	3	93
Out of Balance	-	1	-	-	2	2	4	8	1	-	18
Not Stated & Not Diagnosed	-	9	18	16	1	4	2	15	2	9	76
General Wear	-	-	1	-	-	-	7	12	6	-	26
Miscellaneous	-	2	1	4	2	5	-	2	1	-	12
Totals	7	142	174	293	36	96	165	312	149	100	1,474

Production and Progress board, giving details of how many engines are under repair, the location of each engine, and the number of each type of engine awaiting despatch to Rouen and the Aircraft Depots.

German Aircraft

It was not only British and French (and, towards the end of the war, the American Liberty) engines that were inspected and overhauled at ERS. Engines from captured German aircraft were also sent to Pont de l'Arche for dismantling and examination. RFC, and later RAF, Headquarters sent reports on all captured aircraft to each wing of the service, and to the aircraft depots and the Engine Repair Shops. Below are two of the more concise reports and are included for general interest.

Report on L.V.G. Two-Seater – Allotted RFC No 108

Brought down by Captain McCudden of No. 56 Squadron on 23rd December near Metz [4].
Machine No: 2517.
Dates of Main Spar: 15th November 1917 and 29th November 1917.
Fuselage: 3-ply construction – no internal bracing; yellow.
Main planes and tail unit: Camouflaged dark green and mauve on top surfaces and light blue on lower.
Elevators: One piece as in the Albatros.
Undercarriage: V Struts are of wood, shock absorber is the elastic type.
Tyres: Continental one side and Harburg of Vienna on the other, size 810 x 125 mm.
Engine: 200hp Benz No 22856 with 2 ZH6 type Bosch magnetos Nos 2041280 and 2045005.
Radiator: Honeycomb type.
Main Petrol Tank: Capacity of approximately 46 gallons.
Armament: 1 Spandau gun No 2808, 1917 type firing through the propeller with a direct drive interrupter gear.
Bombs: A bomb-rack was found amongst the wreckage designed to carry eight 24lb bombs. This was probably

situated in the Observer's cockpit.
Ammunition: This consisted of 62 rounds of ordinary infantry type. No other type was found.

General Staff (Intelligence)
G. Barfoot-Saunt 2/Lt RFC, Headquarters, Royal Flying Corps, 3rd January 1918.

Report on Fokker Dr1 Triplane – Allotted RAF No G/5 BDE/2

Details of capture: This machine was flown by Captain Baron Von Richtofen and was shot down by Captain Brown of 209 Squadron in Sopwith Camel No. 7370 on 22nd April 1918 near Corbie [5].

General: It is a complete wreck and was exposed to shell-fire for some hours, but as far as could be ascertained the construction of the machine appears to be similar to G125 (Leutnant Von Stapenhorst's Fokker Dr1 144/17 brought down by anti-aircraft fire on 3rd March 1918). It is painted bright red all over.

Date on top plane: 13th December 1917.
Machine No: 2009
Fabric: This appears to be of rather better quality than usual.
Tyres: Continental size 760 x 100 mm.
Shock Absorber on the undercarriage: The elastic strand type.
Engine: This is a copy of the 110 hp Le Rhone by the Oberursel Company, No 2478, fitted with Bosch magneto No 2202160 type ZH6. The finish of the engine is better than those captured in previous machines of this type.
Guns: Two Spandau guns, No 1795 and 659, firing through the propeller, actuated by a direct flexible drive interrupter gear.
General Staff (Intelligence)

G Barfoot-Saunt 2/Lt RAF, Headquarters, Royal Air Force, 25th April 1918.

Maybach Engine

On 24th October 1917, after a thorough examination at ERS, a detailed description of a captured German Maybach Engine was circulated throughout the Royal Flying Corps [the Maybach powered the Zeppelin Staaken four-engined giant bombers designed to bomb major cities in England]:

General Description: 6 cylinder vertical water-cooled, the cylinders mounted separately.
Bore: 165 mm.
Stroke: 180 mm.
Weight: (without magnetos, carburettor, water and oil pumps) 765 lbs.
Sparking plugs: Two mounted vertically in each cylinder head, one at the front and the other at the rear.
Valves: One pair of exhaust and one pair of inlet valves on opposite sides of each cylinder head.
Diameter of valve seating: 49 mm.
Camshafts: There are two, one working the exhaust valves, the other on the other side of the crankcase working the inlet valves.
Pistons: Cast iron, with four piston rings and no scraper ring.
Small end: 44 mm diameter by 91 mm in length.
Big ends: Brass shells with white metal linings, 66 mm diameter by 73 mm in length.
Crankcase: Aluminium.
Propeller: Axial – Diameter 3200 mm, Pitch 1800 mm.

1918 Armistice and Beyond

In January 1918, Acting Major David Samuel Jillings M.C. joined the ERS as adjutant. Born in 1881, he enlisted in 1899; as a Lance-Sergeant in 1905 he passed a cookery course, and in 1907 he passed a musketry course. He transferred to the RFC in 1913, and has the distinction, when a Sergeant Major, of being the very first RFC casualty to be caused by enemy action; during a reconnaissance flight on the 22nd August 1914, as observer with No. 2 Squadron, he was wounded in the leg by a rifle bullet. After recovering from his injury, he was promoted and qualified as a pilot serving with No. 14 Squadron before eventually being posted to ERS. He retired from the RAF in 1926 as Squadron Leader D. S. Jillings M.C.

Captains R. H. Verney, G. P. Bulman, and F. Halford of the Air Inspection Department made frequent visits to discuss the overhaul of engines and to discover the latest ERS innovations and methods of engine repair. The Engine Repair Shops had a good relationship with the AID. This was helped, as referred to earlier, by the fact that Captain Verney, who succeeded Captain R. K. Bagnall Wild as Inspector of Engines, had joined the AID from ERS late in 1915, so had a good knowledge of the work carried out at Pont de l'Arche.

Towards the latter period of the war, the emphasis for manufacturers was the production of high-powered engines for the latest fighting aircraft; however, flying training schools were based around the Avro 504J, which was powered by the relatively low-powered Gnome Monosoupape rotary engine. The only way Avro could produce the required number of completed 504s was if ERS in France could provide the necessary quantity of reconditioned engines. In late 1917, to help facilitate the provision of these engines, Roy Chadwick, draughtsman and designer from A. V. Roe, visited Pont de l'Arche. Consequently, a large number of Le Rhone and Clerget engines were made available, and to accommodate the different engines a new mounting was conceived by the manufacturers. Aircraft with this new engine mounting were designated Avro 504K. Roy Chadwick later described ERS as *"a veritable cathedral dedicated to aero engines"* (Penrose, 1985).

Avro 504K D7623, one of a batch of 300 Avro 504Js and 504Ks built by A. V. Roe (FlightGlobal).

The War Office (the army) and the Admiralty (the navy) had been competing with each other for equipment and resources to fund their respective flying services, which caused severe problems with supply. British Prime Minister, David Lloyd George, appointed General Jan Christian Smuts to solve the problem. Smuts's recommendations resulted in the amalgamation of the Royal Flying Corps and the Royal Naval Air Service. Thus on 1st April 1918, the first air force to be independent of either army or navy was born: the Royal Air Force.

On the morning of 11th November 1918, all ERS personnel were assembled on the parade ground for the announcement of the signing of the Armistice. Storeswoman Winifred Wilcox recalls standing to attention with colleagues and German prisoners of war, and just weeping. Elsewhere in the camp, however, there were immediate celebrations; some of the mechanics, including Thomas Boland, went up to the railway line and flagged down the first train to Rouen, where they joined in the many celebrations there, before returning to ERS. Before the end of the day, Winifred Wilcox and Arthur Mandeville had become engaged.

At the time of the Armistice, the Engine Repair Shops was still very much a 'going concern', and both Colonel Hynes and Lieutenant Colonel Fell had been awarded the Distinguished Service Order (DSO). No home leave was available, as every boat was full of troops being demobilised. Thomas Boland and some of his colleagues were fortunate to spend a ten-day leave with the padre in Lourdes. Everyone was able to enjoy a sumptuous Christmas dinner back at camp, which included pork, turkey and all the trimmings.

Officers of the Engine Repair Shops 1918. Seated in the centre of the second row are Lt. Col. L. F. R. Fell (wearing forage cap), Colonel G. B. Hynes, Major J. Starling (H. Fell).

In January 1919, Fell was awarded the OBE (Order of the British Empire). Meanwhile, work was gradually being wound down at ERS; such a large establishment was no longer required in peacetime. Not long before leaving for England, a group photograph was taken of the remaining air mechanics, including Thomas Boland, one of whom is holding a sign which reads: *"All that is left of 'em"*. Not all personnel returned to Britain, however. During their time at the Engine Repair Shops, some mechanics had become so friendly with local families (or the daughters of local families!), that they decided to stay in the area. William Delauney, Alfred Turvey and Sidney Warren, among others, decided to stay in Pont de l'Arche or Les Damps; in contrast, mechanic Harry Ainsworth married local girl Madeleine Duparc and they made their home in England.

Winifred Wilcox and Arthur Mandeville married in Steyning, West Sussex, living in Seaford where Arthur worked for the Southdown Motor Services Bus Company. They had three children before Arthur died in 1942; Winifred remained in Seaford until her death in 1991.

Thomas Boland returned to Keighley, where, in 1922, he married local girl Ada Grainger and together they brought up four children, two boys and two girls born between 1922 and 1931.

Colonel Hynes DSO received a permanent commission in the RAF, and by 1921 was the Chief Experimental Officer (engines) at the Royal Aircraft Establishment, Farnborough. In 1923 he became Principal Technical Officer before moving to the Air Inspection Department, eventually in the role of Deputy Director of Aeronautical Research with the rank of Group Captain RAF.

"All that is left of 'em". Thomas Boland is seated on the front row, second from the right, wearing football boots.

In 1920, Lieutenant Colonel Louis Frederick Rudston Fell DSO OBE married Mary Dolores Maud Hayne Walker, known to everyone as Mollie, in Knaresborough, Yorkshire, before moving to the village of North Cray in Kent, from where Fell could commute to his London office. They had first met at the Engine Repair Shops where Mollie served as a driver.

No 46 Reserve Squadron's Crossley 20/25 hp Tender outside Tadcaster Post Office in North Yorkshire in 1917. This type of vehicle was driven by Mollie Walker at Pont de l'Arche. More than six thousand Crossley Tenders were supplied to RFC and RAF units during the war (J. Payne).

Although only just over 5ft tall, Mollie was an excellent driver despite having to contend with the vagaries of the ubiquitous Crossley Tender (general service light truck), with which every RFC and RAF unit was equipped. Mollie even sang about the Crossley in the 9th November show at the 'Pontodrome', where she was one of the main performers (see the programme for the show in the Appendix). They later had two sons, Henry, who became a successful sheep breeder and farmer, and John, who became an engineer and journalist.

Mollie Walker. Copy of a photograph from one of the shows at the 'Pontodrome' (H. Fell).

Lt. Col. Fell became Chief of Aero Engine Design and Research at the Air Ministry, where he drew up the first schedule of tests to establish a general standard of engine reliability. He also designed a two-stroke engine specifically intended to give low fuel consumption. In 1928, he joined Rolls Royce as Technical Assistant to the Managing Director, where he was instrumental in persuading Henry Royce to design an engine for the 1929 Schneider Trophy Contest. This engine was eventually developed to become the Rolls Royce Merlin engine which powered the Spitfire, Hurricane, Lancaster, and many more types of aircraft during the Second World War. From 1934 until 1939, Colonel Fell was Armstrong Siddeley's Chief Engineer, before returning to Rolls Royce at Derby. As Chief Power Plant Engineer, Colonel Fell, with Roy Chadwick (who designed most of Avro's successful aircraft), was responsible for the design and supply of engine installations for the conversion of the twin-engined Avro Manchester aircraft to the four-engined Avro Lancaster (their first meeting had been in 1918 when Chadwick visited ERS regarding the supply of rotary engines for the Avro 504). During the period that Lt. Col. Fell was making his mark in engineering, his younger sister, Dame Honor Bridget Fell DBE FRS, was forging her own career in cell biology and arthritis research.

Following the end of the Second World War, Colonel Fell became Technical Sales Manager, then Public Relations Manager, before ending his Rolls Royce career as consultant on railway traction. During this time he patented two transmission systems for British Railways, one used in the 'Fell' 4-8-4 experimental diesel locomotive no. 10100, and the other in the Yorkshire Engine Company's 'Taurus' locomotive.

Lt. Col. Fell (wearing glasses) discussing the controls of the 'Fell' Locomotive (H. Fell).

Postscript

On the 21st December 1918, the Commander of the Royal Air Force, Major General John Salmond, wrote in a letter to Colonel Hynes:

> *Colonel Hynes, D.S.O.*
> *o/c Engine Repair Shops,*
> *Pont de l'Arche.*
>
> *"Now that demobilisation is approaching, I wish once more to express to you and all ranks of the Engine Repair Shops, my appreciation of the work that you have done for the Royal Air Force. Your work throughout has been of the highest standard, and in times of stress you have never failed to respond cheerfully to the special demands made on you. You have established a reputation for excellence of workmanship not only in the Royal Air Force, but also amongst the flying press of the allied armies, and I feel sure that for all future years each one of you will be proud to have served in the Engine Repair Shops of the RAF in France."*
>
> *J. M. Salmond,*
> *Major General,*
> *Commanding Royal Air Force,*
> *In the Field.*
>
> *21st December 1918.*

.

Endnotes

1. American Samuel Franklin Cody (1867-1913) starred in Wild West shows in theatres around Britain before becoming interested in kites, particularly kites that were big enough to take the weight of a man. Employed by the British army, he then progressed from kites to gliders to powered flight. He was killed in 1913, when his Cody aircraft crashed into trees.

2. Lieutenant General Sir David Henderson KCB, KCVO, DSO (1862-1921) served with the British army in Zululand, Ceylon (now Sri Lanka), Sudan, and the Boer War. In 1911, he qualified as a pilot and at the outbreak of war was in charge of the RFC in the field.

3. The National Physical Laboratory had been conducting investigations into light alloys for many years, and during the war period much experimental work was carried out at the Royal Aircraft Factory, the Universities of Manchester and Birmingham, and at works factories, as well as by the Aeronautical Inspection Department. A large amount of valuable information was thus made available to manufacturers of engines and aircraft parts and also to the Engine Repair Shops.

4. Major James Thomas Byford McCudden VC, DSO & bar, MC & bar, MM, Croix de Guerre (1895-1918) transferred from the Royal Engineers to the RFC as a mechanic, progressing through the ranks until by the time of his death in a flying accident in 1918, he was one of the highest-scoring fighter aces of the war. This L.V.G. reconnaissance aircraft was one of four German machines he shot down on 23rd December 1917.

5. Rittmeister Baron Manfred von Richtofen (1892-1918) was the highest scoring ace of the war with 80 victories. He was shot down and killed on the 21st April 1918 (not the 22nd April as the report states). Although Canadian Sopwith Camel pilot Roy Brown was credited with the victory, more recent research suggests that Richtofen was actually brought down by groundfire.

Appendix 1

Engine Data

The engines in the list below provided the power for most of the aircraft operated by the RFC, RNAS and RAF during the First World War. As can be seen, the majority of these engines were utilised by various aircraft types; conversely the same aircraft type could be powered by more than one make of engine. This list is not exhaustive.

80 hp Gnome
Avro 504, Bristol Scout, B.E.8, Sopwith Pup, Morane-Saulnier Monoplane.
> Air-cooled rotary engine
> 7 cylinders
> 1150 rpm
> 124 mm bore
> 140 mm stroke
> Oil consumption: 1½-1¾ gallons/hour
> Petrol consumption: 7-8 gallons/hour
> Weight: 210 lbs

80 hp Le Rhone
Avro 504, Bristol Scout, Sopwith Pup, Morane-Saulnier Monoplane, Nieuport 11.
> Air-cooled rotary engine
> 9 cylinders
> 1175 rpm
> 105 mm bore
> 140 mm stroke
> Oil consumption: 1 gallon/hour
> Petrol consumption: 6-7 gallons/hour
> Weight: 240 lbs

110 hp Le Rhone
D.H.5, Bristol M1, Nieuport 17, F.E.8, Sopwith 1½ Strutter, Sopwith Camel.
>Air-cooled rotary engine
>9 cylinders
>1200 rpm
>112 mm bore
>170 mm stroke
>Petrol consumption: 10-11 gallons/hour
>Weight: 330 lbs

130 hp Clerget
Nieuport 17, Sopwith Triplane, Sopwith Camel.
>Air-cooled rotary engine
>9 cylinders
>1250 rpm
>120 mm bore
>160 mm stroke
>Oil consumption: 1¾ gallons/hour
>Petrol consumption: 11¾ gallons/hour
>Weight: 370 lbs

150 hp Bentley BR1
Sopwith Camel.
>Air-cooled rotary
>9 cylinders
>1250 rpm
>120 bore
>170 stroke
>Oil consumption: 1½ gallons/hour
>Petrol consumption: 11 gallons/hour
>Weight: 400 lbs

90 hp R.A.F.1a
Armstrong Whitworth F.K.3, B.E.2c, B.E.2d, B.E.2e.
    Air-cooled 'V' engine
    8 cylinders
    1800 rpm
    100 mm bore
    140 mm stroke
    Oil consumption: 4-5 pints/hour
    Petrol consumption: 8-9 gallons/hour
    Weight: 440 lbs

120 hp Beardmore
Armstrong Whitworth F.K.8, Martinsyde G100, F.E.2b.
    Water-cooled in-line engine
    6 cylinders
    1200 rpm
    130 mm bore
    175 mm stroke
    Oil consumption: 3½ pints/hour
    Petrol consumption: 9½ gallons/hour
    Weight: 630 lbs (incl. radiator and water)

150 hp R.A.F.4a
R.E.7, R.E.8, B.E.12.
    Air-cooled 'V' engine
    12 cylinders
    1800 rpm
    100 mm bore
    140 mm stroke
    Weight: 680 lbs

150 hp Hispano Suiza
S.P.A.D. VII, S.E.5a.
> Water-cooled 'V' engine
> 8 cylinders
> 1500 rpm
> 120 bore
> 130 stroke
> Oil consumption: 6 pints/hour
> Petrol consumption: 13-15 gallons/hour
> Weight: 470 lbs

250 hp Rolls Royce Eagle
D.H.4, Handley Page 0/100, Blackburn Kangaroo, Short Bomber, F.E.2d.
> Water-cooled 'V' engine
> 12 cylinders
> 1800 rpm
> 115 mm bore
> 165 mm stroke
> Oil consumption: 6 pints/hour
> Petrol consumption: 20 gallons/hour
> Weight: 900 lbs

220 hp Rolls Royce Falcon
Bristol F2b.
> Water-cooled 'V' engine
> 12 cylinders
> 2000 rpm
> 102 mm bore
> 146 mm stroke
> Oil consumption: 6 pints/hour
> Petrol consumption: 18-20 gallons/hour
> Weight: 695 lbs

250 hp Siddeley Puma
D.H.9.
> Water-cooled in-line engine
> 6 cylinders
> 1400 rpm
> 145 mm bore
> 190 mm stroke
> Oil consumption: 1¼ gallons/hour
> Petrol consumption: 18¾ gallons/hour
> Weight: 625 lbs

400 hp Liberty Engine
D.H.9A, American D.H.4
> Water-cooled 'V' engine
> 12 cylinders
> 1750 rpm
> 127 mm bore
> 178 mm stroke
> Oil Consumption: 1½ gallons/hour
> Petrol consumption: 30 gallons/hour
> Weight: 820 lbs

Appendix 2

U.S. Patent Application – Lieutenant Colonel L. F. R. Fell – 31st July 1918

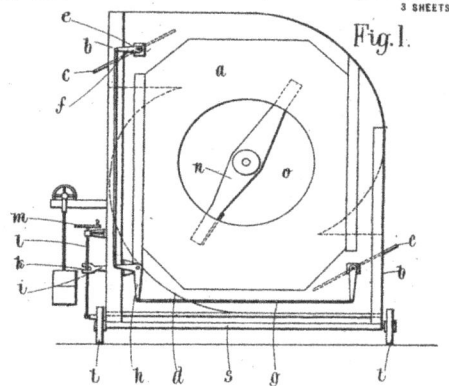

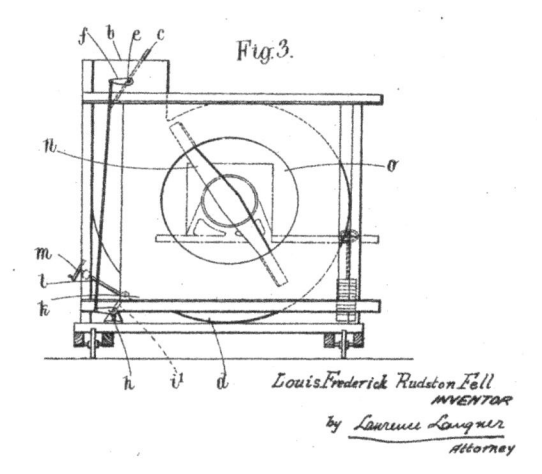

Means for Absorbing Power and Determining the Reaction on Engines
Patented 22nd February 1921

L. F. R. FELL.
MEANS FOR ABSORBING POWER AND DETERMINING THE REACTION ON ENGINES.
APPLICATION FILED JULY 31, 1918.

1,369,018.

Patented Feb. 22, 1921.

3 SHEETS—SHEET 2.

Fig. 2.

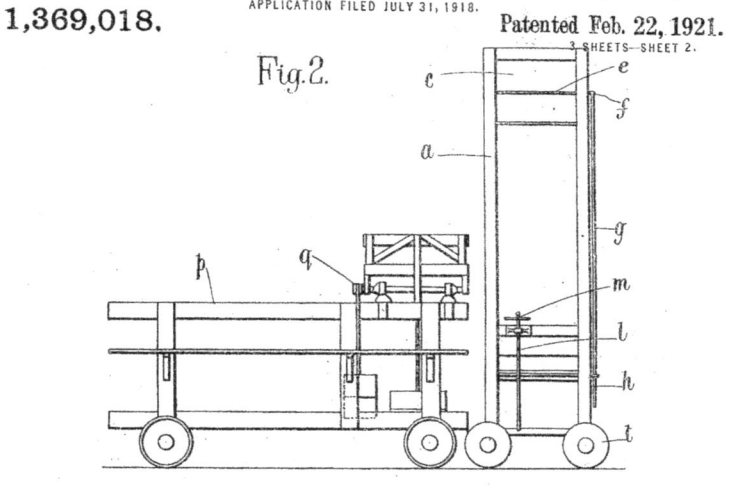

Fig. 4.

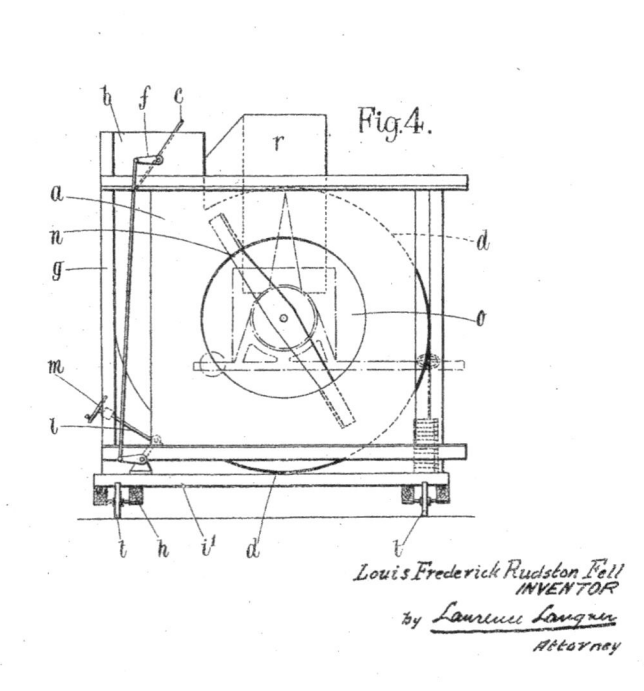

Louis Frederick Rudston Fell
INVENTOR
by Lawrence Langner
Attorney

L. F. R. FELL.
MEANS FOR ABSORBING POWER AND DETERMINING THE REACTION ON ENGINES.
APPLICATION FILED JULY 31, 1918.

1,369,018.

Patented Feb. 22, 1921.
3 SHEETS—SHEET 3.

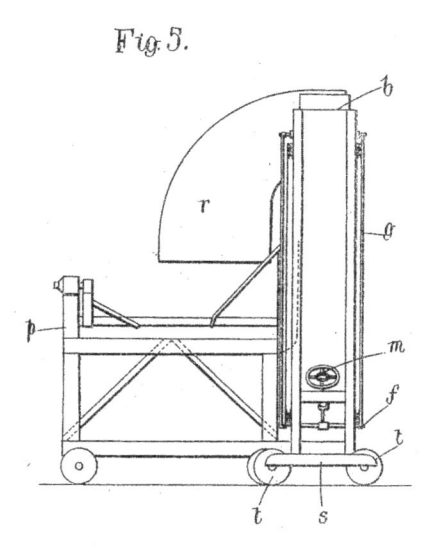

Fig. 3.

Louis Frederick Rudston Fell
INVENTOR
by Lawrence Langner
Attorney

UNITED STATES PATENT OFFICE.

LOUIS FREDERICK RUDSTON FELL, OF FOWTHORPE, FILEY, ENGLAND.

MEANS FOR ABSORBING POWER AND DETERMINING THE REACTION ON ENGINES.

1,369,018.   Specification of Letters Patent.   Patented Feb. 22, 1921.

Application filed July 31, 1918.   Serial No. 247,695.

To all whom it may concern:

Be it known that I, LOUIS FREDERICK RUDSTON FELL, a subject of the King of Great Britain, residing at Fowthorpe, Filey, England, in the county of Yorkshire, have invented new and useful Improvements in and Relating to Means for Absorbing Power and Determining the Reaction on Engines, of which the following is a specification.

This invention has reference to absorbing power and determining the reaction on engines more particularly when the engine is run at different speeds and developing different horse powers. The object of this invention is to determine the amount of reaction exerted on the engine bearers, especially aero engine bearers, with variations in engine load.

The invention therefore consists in absorbing the power developed by an engine at various speeds by a controllable flow of air and measuring the reaction of the absorbed power on said engine.

This invention contemplates a variable air brake for absorbing the power developed by the engine to be tested in combination with an engine support responsive to variations in torque reaction due to variations in load on the engine and means for measuring that reaction.

In order that the invention may be readily understood and carried into effect same will now be described more fully with reference to the accompanying drawings in which:—

Figure 1 is an elevation of an embodiment of the invention in which provision is made in the device for two valved outlets for air.

Fig. 2 is a side view with the test bench and engine bearer adjacent thereto.

Fig. 3 is an elevation illustrating an arrangement involving a single valved outlet for air.

Fig. 4 is an elevation and Fig. 5 a side view illustrating a further arrangement in which air set in motion by the engine is utilized in cooling the engine.

According to the embodiment illustrated in Figs. 1 and 2 a casing a which may be constructed of wood built up on a suitable framework and lined with sheet metal is provided having suitably disposed air outlets b b whereof the effective area is controlled by appropriate valves c c. The casing incloses a chamber d preferably of snail like form as shown by the dotted lines from which the outlets b b extend the said chamber being conveniently formed by the sheet metal lining or composed of wood slats closely arranged and secured so as to render the walls of the chamber approximately air tight. The valves c may be operated by mechanism under a single control or each valve may be controlled separately. Preferably the valves are connected so as to be simultaneously opened or closed either manually, mechanically or otherwise. In the arrangement shown each valve c is of butterfly form and consists of wings or flaps mounted on a shaft or spindle e to each of which latter a lever f is secured. These levers f are each connected by a rod g with a bell-crank lever h which carries an extension i provided with a nut k adapted to be operated by the screw spindle l to which the operating wheel m or a suitable handle is applied. Mounted in the aforesaid chamber is a suitable fan-brake n adapted for being secured to the crank shaft of the engine to be tested so as to be driven thereby or to be driven through appropriate intermediate gearing as may be found convenient according to the type of engine under test. The said fan-brake when rotated in the chamber of the casing causes air which is admitted thereto in any suitable manner such as by means of the aperture or apertures o therein through which the crank shaft of the engine passes to flow through the said chamber and through the aforesaid outlets b b.

The casing may be formed with suitable doors on one or both sides to afford access to the interior of the casing and the fan-brake if and when required.

In constructing the fan-brake the dimensions are calculated or determined by appropriate formulæ in such a manner as to render it capable of absorbing the maximum power which the engine under test is likely to develop at the lowest number of revolutions per minute at which it is desired to run.

In operation the improved method and apparatus are employed in conjunction with an ordinary balance bar torque reaction test bench indicated at p the power absorbed by the fan-brake being adjusted that is to say decreased or increased by either closing or opening more or less the aforesaid valves c c the reaction of the power absorbed by the fan-brake being calculated from the balance bar q of the torque bench in the usual manner by the standard formulæ. Thus the air which is drawn into the casing by the action of the fan is caused to flow from the said casing in a regulated manner with the result that the power developed by the engine and absorbed by the fan-brake reacts upon the engine under test and this reaction is calculable upon the balance bar pertaining to the torque bench.

For use with the aforesaid apparatus a test bench whereof the engine bearers on the cradle are adapted for being moved relatively to the axis of the fan-brake has been found most convenient as such arrangement renders the test bench capable of adaptation to any type of aero engine.

A construction in which a single controlled outlet for air is employed is illustrated in Fig. 3. The various parts of the apparatus bear reference letters similar to the respective parts in Figs. 1 and 2 and therefore the operation will be clearly understood without detailed description thereof. This arrangement is serviceable in cases where the test bench is so arranged that the torque of the engine tends to effect the rotation of the engine about the axis of the fan-brake and in connection therewith it will be seen that the valve actuating wheel m and screw spindle l operate directly upon a nut k carried by the arm i^1 operatively connected with the lever h.

The arrangement of valve actuating levers and rods may be duplicated one set being disposed on each side of the fan casing.

If an air cooled engine be under test a part of the air set in motion by the operation of the engine may be diverted so as to serve in cooling the cylinder heads. An arrangement of the apparatus adapted for accomplishing this object is shown in Figs. 4 and 5 a being the casing b the outlet with controlling valve c and d the snail form of air chamber. A flue or conduit r is arranged in communication with the chamber d and outlet b the said conduit being appropriately formed or curved as seen in Fig. 5 so as to direct the diverted air from the said chamber d on to the cylinder head of the engine (not shown) which is supported by the test bench p.

For convenience of transport the apparatus may be mounted upon a carriage s supported by running wheels t. Thus it may be moved from place to place with ease as required.

The improved means for testing is especially adapted for use with aero engines as the strains to which the crank shaft of the engine under test is submitted are only those which will be borne by the said shaft during the operative service of the engine, the stress ordinarily sustained by the shaft from revolving a rotor in an inelastic medium such as water when a water brake is used being avoided.

What I claim and desire to secure by Letters Patent of the United States is:—

1. The combination with a fan brake and means for regulating the volume of air passing therethrough for absorbing the power developed by an engine, of an engine bench rotatable about the axis of the fan brake and responsive to variation in torque reaction, due to the power absorbed, and means for measuring that reaction.

2. The combination with an engine, of a fan coupled thereto, a snail-like casing in operative relationship to said fan, a central air inlet opening in said casing, peripheral air outlets in said casing, valves for controlling the flow of air through said outlets to regulate the power absorbed by said fan, an engine support rotatable about the axis of said fan, means for securing the engine to said support and a balance bar mounted on said engine support whereby the torque reaction due to the power absorbed is measured.

3. The combination with a fan coupled to an engine under test, of a casing for said fan, means for regulating the volume of air passing through said casing for absorbing the power developed by the engine a conduit appropriately formed with a depending outlet adapted to conduct the diverted air from said casing on to the cylinder head of the engine, a movable support for the engine responsive to variations in the torque reaction due to absorbed power and means mounted on the engine support for measuring that reaction.

LOUIS FREDERICK RUDSTON FELL.

Appendix 3

ERS Athletics Sports Programme

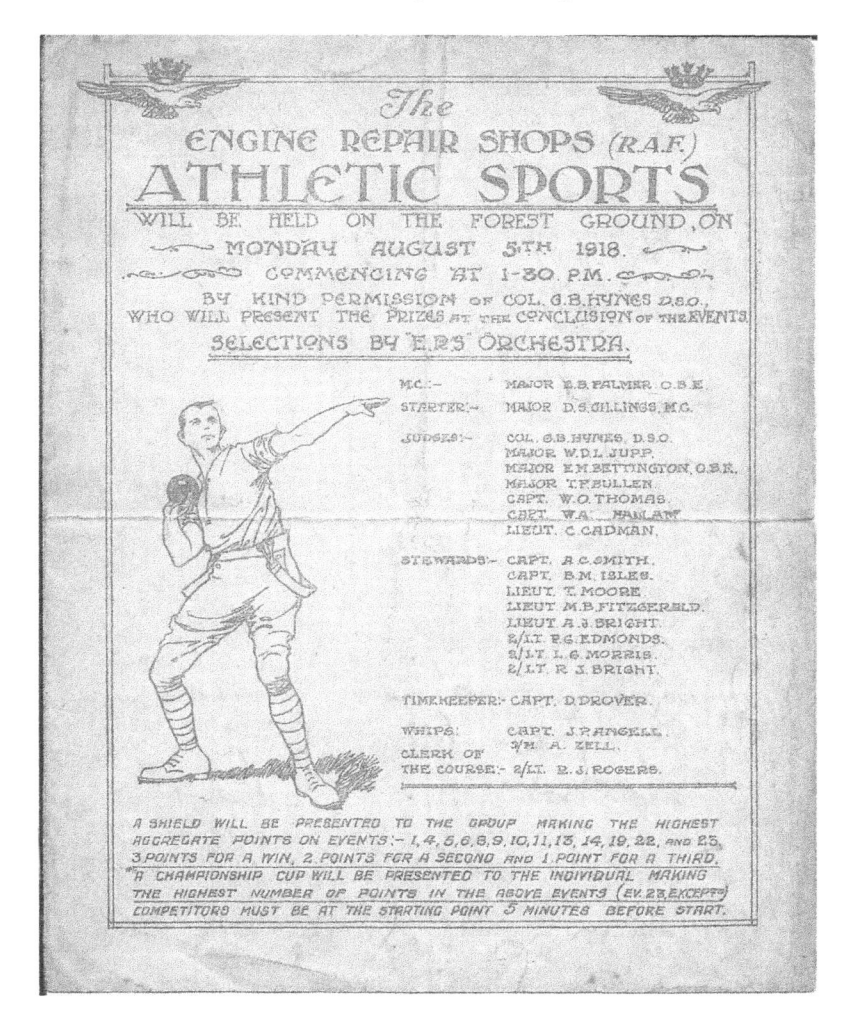

The Starter was Major D. S. Jillings MC (misspelled Gillings on the programme cover). He joined ERS in early 1918; in August 1914, he was the first RFC casualty to be caused by enemy action when he was shot in the leg during a reconnaissance flight.

PROGR[AMME]

	EVENT	TIME	GROUP & INDIV. PTS.	NO. OF PRIZES
1.	PUTTING THE SHOT, SIX FOOT CIRCLE, THREE THROWS.	1-30 P.M.	YES	2
2.	GROUP 100 YARDS FLAT, TWO HEATS, FIRST THREE IN EACH HEAT TO RUN IN FINAL. (AIR MECHANICS)	1-50 P.M.		
3.	DO. (N.C.O.'S)	1-55 P.M.		
4.	FINAL OF EVENT 2.	2-0 P.M.	YES	3
5.	FINAL OF EVENT 3.	2-5 P.M.	YES	3
6.	HIGH JUMP	2-10 P.M.	YES	3
7.	Q.M.A.A.C. SACK RACE	2-30 P.M.		3
8.	100 YARDS, FLAT TWO HEATS & FINAL. (ALL RANKS)	2-35 P.M.	YES	3
9.	HALF MILE, OPEN TO E.R.S	2-40 P.M.	YES	3
10.	LONG JUMP, " "	2-45 P.M.	YES	3
11.	¼ MILE FLAT, (FINAL)	3-5 P.M.	YES	3
12.	TILTING THE BUCKET. OPEN TO SIX PAIRS FROM EACH GROUP	3-10 P.M.		4

A close look at the content of the programme reveals some unusual sporting events, such as 'tilting the bucket'. The members of the QMAAC had their own sack race.

RAMME

	EVENT	TIME	GROUP & INDIV. PTS.	NO. OF PRIZES
13.	RELAY RACE, TEAM OF 4 PER GROUP 220 YDS, ¼ MILE, 220 YDS, ½ MILE.	3-30 P.M.	YES	3
14.	POLE JUMP	3-40 P.M.	YES	3
15.	MUSICAL CHAIRS (CYCLES) OPEN TO Q.M.A.A.C. & E·R·S	3-50 P.M.		3
16.	BOLSTER BAR, OPEN TO E·R·S	4-5 P.M.		4
17.	TUG OF WAR, TEN TO PULL, ANY WEIGHT. HEATS:— NO. 6 GROUP v. NO. 2 GROUP "A" OFFICERS v. NO. 5 GROUP "B" NO. 3 GROUP v. W.O's & SGTS. "C" NO. 1 GROUP v. NO. 4 GROUP "D" SEMI-FINAL WINNERS, A. v B & C v D	4-35 P.M.		
18.	THREE LEGGED RACE (PAIRS 1 R·A·F & 1 Q.M.A.A.C.) POST ENTRIES	5-10 P.M.		3
19.	MILE RACE, OPEN TO E·R·S	5-15 P.M.	YES	3
20.	OFFICERS SACK RACE	5-25 P.M.		3
21.	RELAY RACE. CANADIAN FORESTRY CORPS v. E·R·S (SAME AS EVENT 13)	5-35 P.M.		1
22.	OBSTACLE RACE	5-55 P.M.	YES	4
23.	TUG OF WAR (FINAL) WINNERS OF "A" & "B" v. WINNERS OF "B" & "C"	6-10 P.M.	YES, GROUP ONLY	2
24.	BAND RACE (SPRINT)			3

Interesting events include 'musical chairs', a 'three-legged race', and also a relay race against the Canadian Forestry Corps (a corps of the Canadian Army formed in November 1916 to provide lumber, and to clear areas of land, for use by the Allied forces).

Programme of Music

March.	Entry of the Bulgars.	Lotter.
Overture.	Poet and Peasant.	Suppé.
Valse.	Nights of Gladness.	Ancliffe.
Selection.	A Country Girl.	Monckton.

— INTERVAL —

March.	Viscount Nelson.	Zehle.
Overture.	Light Cavalry.	Suppé.
Valse.	Luna.	Lincke.
Selection.	The Geisha.	Jones.

— INTERVAL —

March.	Under the Banner of Victory.	Blon.
Valse.	Santiago.	Corbin.
Selection.	Happy days of Dixie.	Bidgood.
Song.	Somewhere a voice is calling.	Tate.
Galop.	En Avant.	Barthmann.

God Save the King.

Band President:— Major D. Jillings, M.C.

Conductor:— S/m. Williamson.

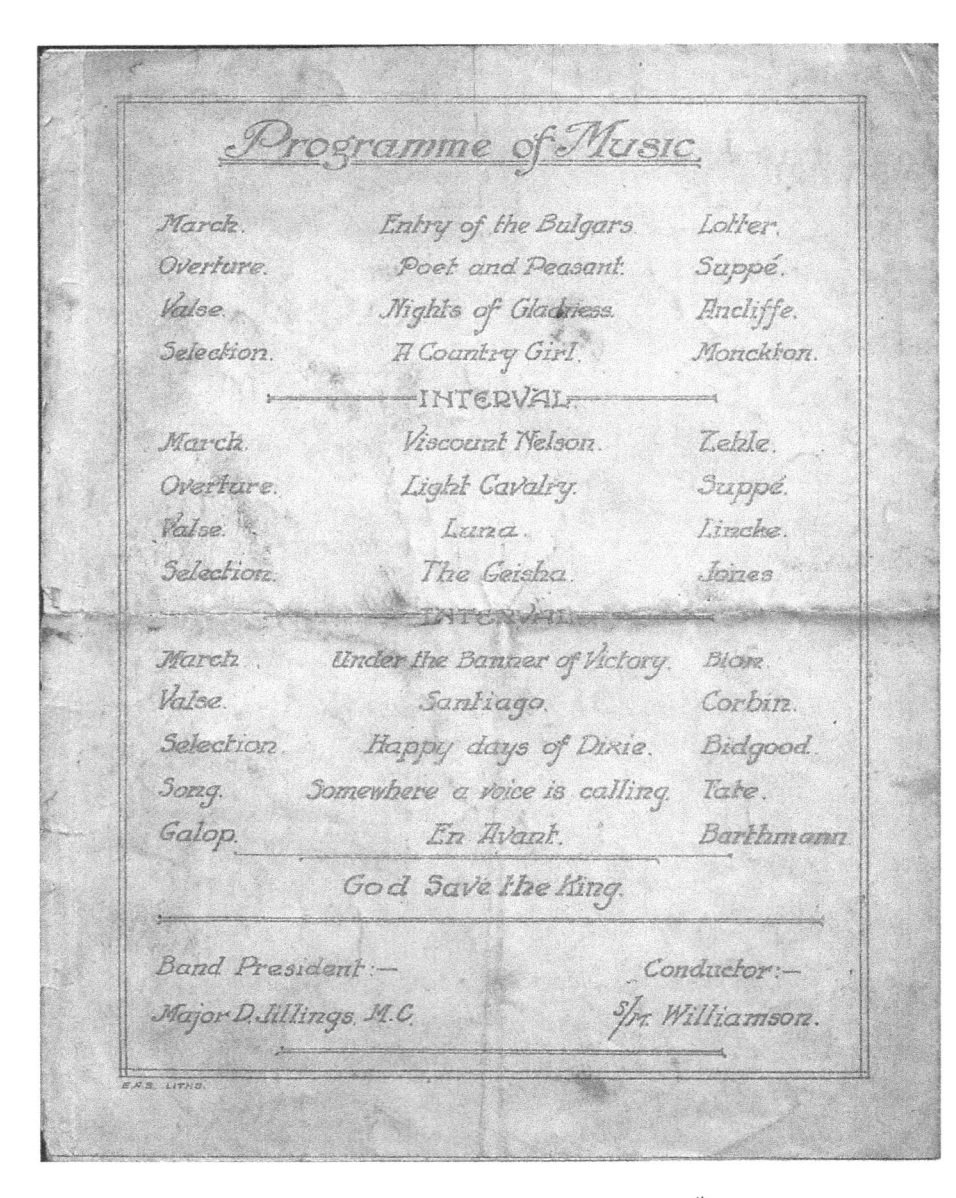

According to a local newspaper report, on this date (5th August 1918) it rained for most of the day, but this did not dampen the spirits of everyone involved in the sporting activities.

100

Appendix 4

'The Astras'

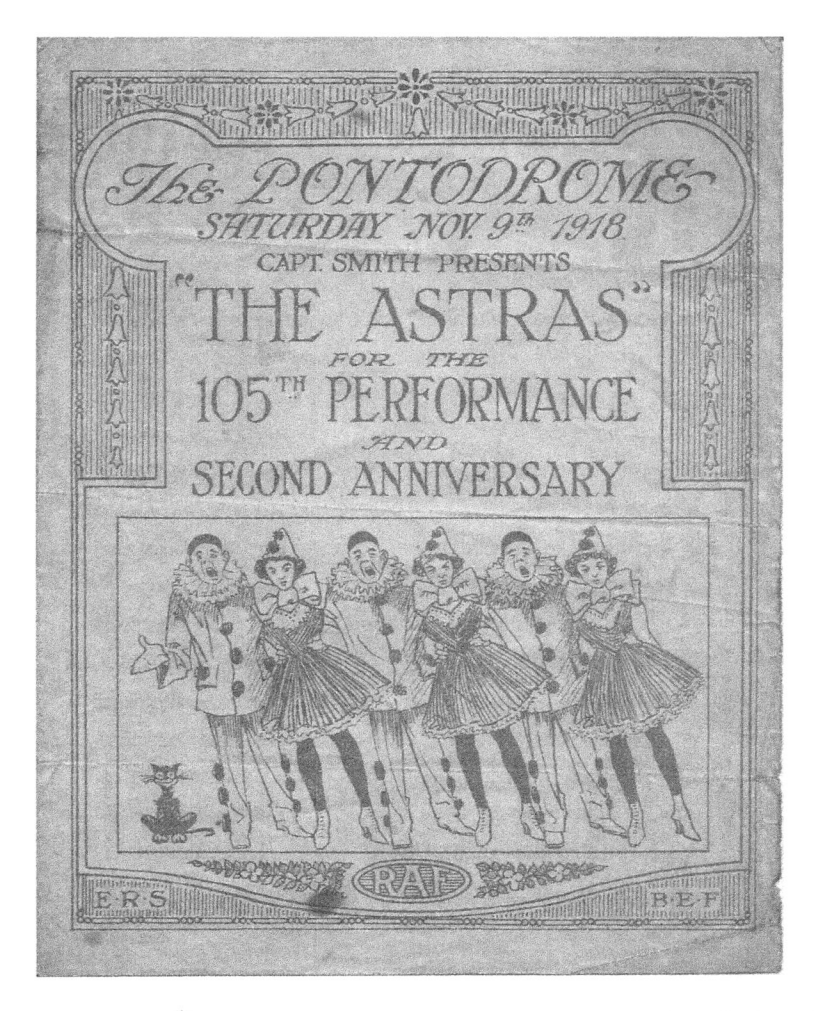

Programme for 9th November 1918 performance (just two days before the Armistice). The motto of the Royal Air Force, and previously the Royal Flying Corps, is *"Per Ardua ad Astra"*, meaning "Through Adversity to the Stars"; hence the performers at the 'Pontodrome' calling themselves the *'Astras'*, the 'Stars'.

PROGR[AMME]

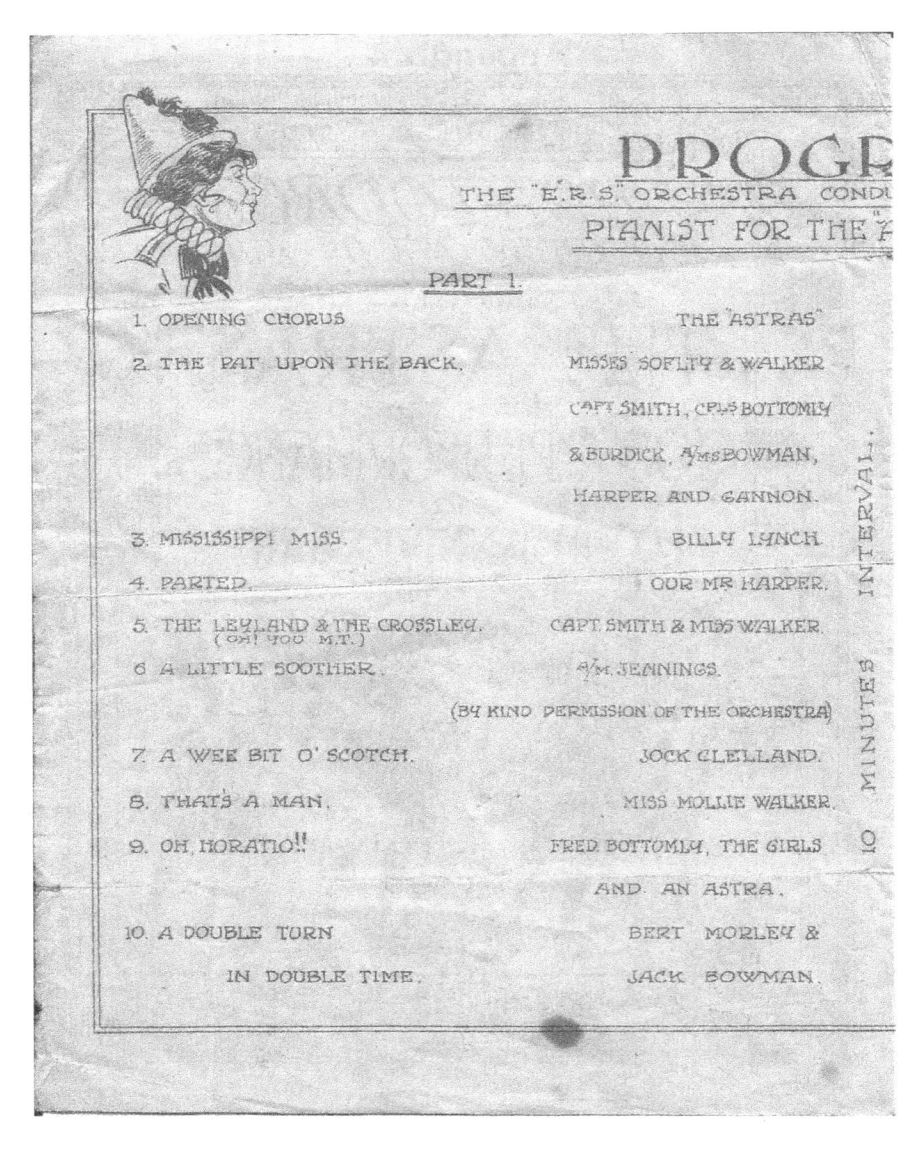

THE "E.R.S." ORCHESTRA COND[UCTED]

PIANIST FOR THE [...]

PART 1.

1. OPENING CHORUS	THE "ASTRAS"
2. THE PAT UPON THE BACK.	MISSES SOFLTY & WALKER
	CAPT. SMITH, CPLS BOTTOMLY
	& BURDICK, A/MS BOWMAN,
	HARPER AND CANNON.
3. MISSISSIPPI MISS.	BILLY LYNCH.
4. PARTED.	OUR MR HARPER.
5. THE LEYLAND & THE CROSSLEY. (OH! YOU M.T.)	CAPT. SMITH & MISS WALKER.
6. A LITTLE SOOTHER.	A/M. JENNINGS.
	(BY KIND PERMISSION OF THE ORCHESTRA)
7. A WEE BIT O' SCOTCH.	JOCK CLELLAND.
8. THATS A MAN.	MISS MOLLIE WALKER.
9. OH, HORATIO!!	FRED. BOTTOMLY, THE GIRLS
	AND AN ASTRA.
10. A DOUBLE TURN	BERT MORLEY &
IN DOUBLE TIME.	JACK BOWMAN.

INTERVAL 10 MINUTES

Miss Mollie Walker's real name was Mary Walker. She drove a Crossley Tender at ERS, as well as singing about the Crossley (see item no.5 above). In 1920, Mollie and Lt. Col. Fell were married.

?AMME
?CTED BY S/M. WILLIAMSON.
STRAS A/M. HARGRAVES.

PART 2.

1. ENCORE HORATIO. — CPL. BOTTOMLY.
2. OUR IDEA OF A PERFECT DAY. — MISSES SOFTLY & WALKER,
   CAPT. SMITH.
   MESSRS BOTTOMLY & HARPER.
3. YOUR OLD FRIEND. — BERT. MORLEY.
4. THE GLORY OF THE SEA. — BARRY HARPER.
5. OCH AYE! — JOCK CLELLAND.
6. LATHAM. — "THE MYSTERIOUS."
7. OH, YOU COON! — JACK BOWMAN.
8. JOAN OF ARC. — MISS DOT SOFTLY.
9. A 'CELLO. AND — FRED JENNINGS.
10. RIGHT ON TOP. — BILLY LYNCH.
11. A YORKSHIRE IDYLL. — MISS SOFTLY & TICH GANNON.
12. THE JAZZ BAND.
    CAPT SMITH, CPL. MORLEY P/Ms. BOWMAN & DRAKE.

GOD SAVE THE KING.

THE ORCHESTRA.

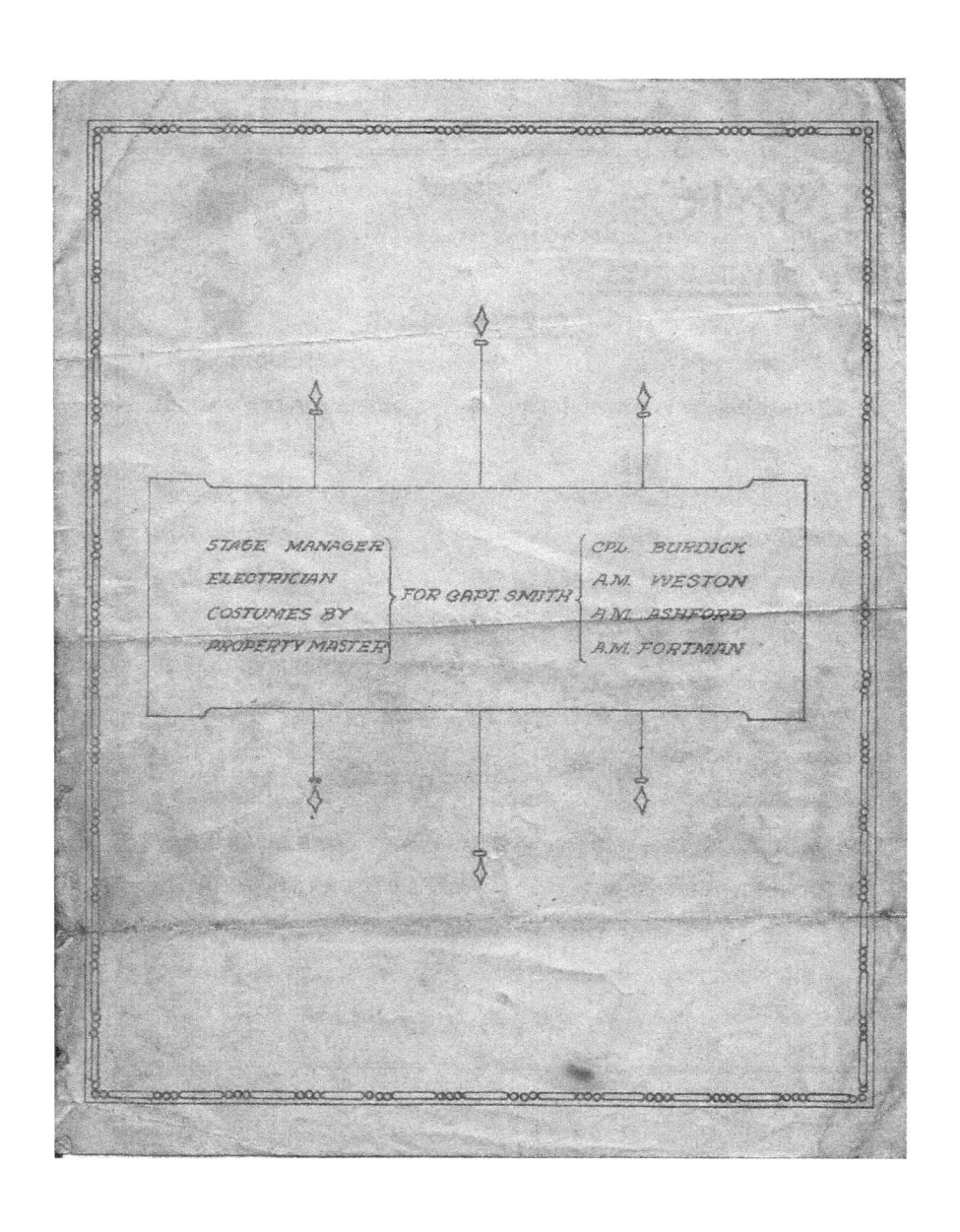

STAGE MANAGER	CPL. BURDICK
ELECTRICIAN	A.M. WESTON
COSTUMES BY	A.M. ASHFORD
PROPERTY MASTER	A.M. FORTMAN

FOR CAPT. SMITH

Copy of a studio portrait of the debonair pre-war professional actor, and producer of the shows at the Engine Repair Shops, Captain A. C. Smith (H. Fell).

The Opening Chorus. Captain Smith in the centre, Mollie Walker at far right (H. Fell).

Appendix 5

Letter of Appreciation
Major General J. M. Salmond 21st December 1918

Colonel Hynes, D.S.O.
%o Engine Repair Shops,
Pont de l'Arche.

Now that demobilisation is approaching, I wish once more to express to you and all ranks of the Engine Repair Shops, my appreciation of the work that you have done for the Royal Air Force. Your work throughout has been of the highest standard, and in times of stress you have never failed to respond cheerfully to the special demands made on you. You have established a reputation for excellence of workmanship not only in the Royal Air Force, but also amongst the flying units of the Allied Armies, and I feel sure that for all future years each one of you will be proud to have served in the Engine Repair Shops of the R.A.F. in France.

J M Salmond
Major General,
Commanding Royal Air Force,
In the Field.

21st. December 1918.

Appendix 6

Christmas Cards

Hand-drawn and painted front page of a Christmas card from the Engine Repair Shops

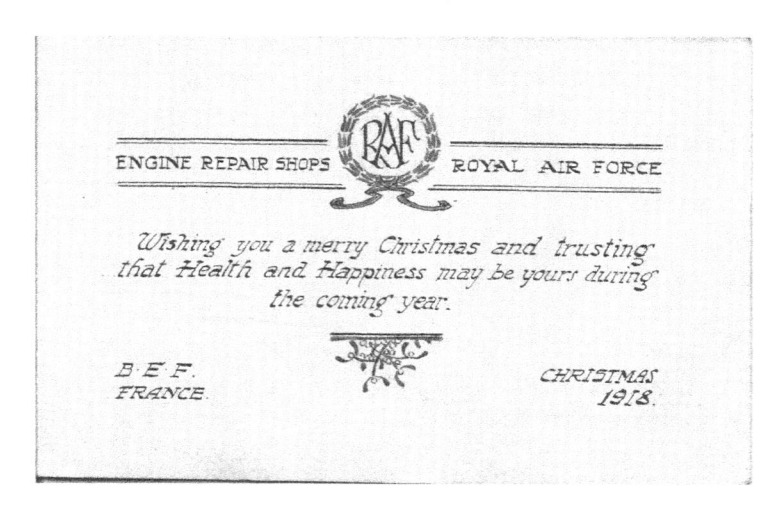

Inside page of the same card (B.E.F. stands for British Expeditionary Force)

Inside page of pen and ink Christmas card

Front page of pen and ink Christmas card

The two Christmas cards, on this and the previous page, are indicative of the high standards of workmanship achieved at Pont de l'Arche.

Acknowledgements

A number of people gave me support, encouragement and advice when I began my research over thirty years ago. My wife, Sheila, allowed me the space and time to put together the original two articles, and again this year during the rewrite. Thanks to my younger daughter, Amy, for proofreading the 2017 manuscript. Thanks to Susan Hutchinson at Lodge Books. Seaford Museum's Chronicler, Kevin Gordon, provided information on WAAC Winifred Wilcox and Air Mechanic Arthur Mandeville. John Sproule (Lt/Cdr RN) (Rt) FRAeS allowed me access to the recording of his interview with Lt. Col. Fell. Henry and Catherine Fell invited me to their home to discuss Henry's father L. F. R. Fell. Air Commodore F. R. (Rod) Banks, CB, OBE, CES, HonFRAes, FIMechE, and Frank Nixon CBE, Bsc, CEng, both shared their memories of Lt. Col. Fell. Neera Puttapipat at the Imperial War Museum, Dave Roberts (RAF Museum), M. H. Evans (Rolls Royce), A. W. L. Naylor (RAeS), Ronald Redman (Narrow Gauge Railway Society), Chaz Bowyer, Barry Gray, Bruce Robertson, Peter Liddle were helpful and answered my queries. Stuart Leslie, Paul Leaman, Dick Barton, Neal Stride, *Cross & Cockade* colleagues, gave advice and support, Kevin Kelly also provided information regarding the purpose-built accommodation at ERS. I visited ex-ERS mechanic Thomas Boland, who shared his memories of ERS, and I also met ex-Sergeant Donald Winn, who shared his memories of 24 Squadron. Mrs Anne Newsome, widow of Captain Thomas H. Newsome, shared his papers and photographs.

Bibliography

Aeronautical Research Committee Reports 1914-1918.
Air Board (1917) *Technical Notes – Engine Notes*. Controller – Technical Department.
Air Board (1917) *Engine Book*. Controller – Technical Department.
Banks, F. R. (1983) *I Kept No Diary*. Shrewsbury: Airlife.
Baring, M. (1920) *RFC HQ 1914-1918*. London: G. Bell & Sons.
Beaumont, R. A. ed (1940) *Aeronautical Engineering*. London: Odhams Press.
Bridgman, L. & Stewart, O. (1972) *The Clouds Remember*. London: Arms & Armour Press.
Bulman, G. P. (1917) AID Report on ERS.
Bulman, G. P. (1966) 'Early Days'. *Journal of the Royal Aeronautical Society*.
Butcher P. E. (1971) *Skill and Devotion*. Hampton Hill: Radio Control Publishing Company.
Edinburgh Gazette – various issues.
Fell, L. F. R. (1966) 'The Engine Repair Shops'. *Journal of the Royal Aeronautical Society*.
Flight – various issues.
FlyPast – various issues.
Gray, P. & Thetford, O. (1962) *German Aircraft of the First World War*. London: Putnam.
Grey, C. G. (1990) *Jane's Fighting Aircraft of World War 1*. London: Studio Editions Ltd.
Gunston, B. (1989) *World Encyclopaedia of Aero Engines*. Cambridge: Patrick Stephens Ltd.
Hare P. H. (1990) *The Royal Aircraft Factory*. London: Putnam.
Heron S. D. (1961) *History of the Aircraft Piston Engine*. Detroit: Ethyl Corporation.

Institution of Royal Engineers (1924) *The Work of the Royal Engineers in the European War 1914-1919 – Work Under The Director of Works (France).* Chatham: W. & J. McKay.
Jackson A. J. (1962) *De Havilland Aircraft since 1909.* London: Putnam.
Jackson A. J. (1965) *Avro Aircraft since 1908.* London: Putnam.
Johnstone, E. G. ed (1931) *Naval Eight.* London: Signal Press Ltd.
Jones, H. A. (1928-1937) *The War in the Air Volumes II to VI.* London: Oxford University Press.
King H. F. (1981) *Sopwith Aircraft 1912-1920.* London: Putnam.
Laumanns, H. W. (2014) *Deutsche Jagdflugzeuge des Ersten Weltkriegs.* Stuttgart: Motorbuch Verlag.
Lewis, C. (1964) *Farewell to Wings.* London: Temple Press Books.
London Gazette – various.
McCosh, F. (1997) *Nissen of the Huts.* B D Publishing.
McCudden, J. T. B (1933) *Flying Fury.* London: John Hamilton.
McInnes, I. & Webb, J. V. (1991) *A Contemptible Little Flying Corps.* London: The London Stamp Exchange Ltd.
Marwick, A. (1977) *Women at War 1914-1918.* Fontana.
Munson, K. (1967) *Aircraft of World War 1.* Shepperton: Ian Allan.
Nahum, A. (1987) *The Rotary Aero Engine.* London: Her Majesty's Stationery Office.
National Archives AIR1/6A/4/36, AIR1/29/15/1/142, AIR1/68/15/9/110, AIR1/160/15/123/7, AIR/697/27/3/1, AIR1/706/27/10/610, AIR1/Box29, AIR1/Box1965, AIR1/Box1967.
Norris, G. (1965) *The Royal Flying Corps a History.* London: Frederick Muller.
Nowarra, H. (1959) *Die Entwicklung der Flugzeuge 1914-1918.* Munich: J. F. Lehmanns Verlag.

O'Connor, M. (2003) *Airfields and Airmen, Cambrai.* Barnsley: Pen & Sword.
Penrose H. (1985) *Architect of Wings.* Shrewsbury: Airlife.
Profile Publications – various.
Raleigh, Sir W. (1922) *The War in the Air Volume I.* London: Oxford University Press.
Shores, C. Franks, N. Guest, R. (1990) *Above the Trenches.* London: Grub Street.
War Office (1968) *RFC Technical Notes 1916.* London: Arms and Armour Press.
Williams, A. (1986) 'The RFC/RAF Engine Repair Shops Pont de l'Arche France'. *Cross and Cockade Journal of the First World War Aviation Historical Society,* vol.17, part 4: pp. 154-161.
Williams, A. (1992) 'The RFC/RAF Engine Repair Shops Pont de l'Arche Revisited'. *Cross and Cockade Journal of the First World War Aviation Historical Society,* vol. 23, part 1: pp. 42-45.

www.airhistory.org.uk
www.airwar1.org.uk
www.aviationhistory.com
www.crossandcockade.com
www.enginehistory.org
www.flightglobal.com/archive
www.greatwar.co.uk
www.iwm.org.uk
www.paxmanhistory.org.uk/fell
www.rafmuseum.org.uk
www.sfcody.org.uk
www.sussexhistory.net
www.thaerodrome.com
www.thevintageaviator.co.nz
www.throughtheireyes2.co.uk
www.wright-brothers.org